W9-ARI-313

Evolution and Human Nature

Also by Richard Morris

The Fate of the Universe
The End of the World
Light

EVOLUTION AND HUMAN NATURE

Richard Morris

SEAVIEW/PUTNAM
New York

Published simultaneously in Canada by
General Publishing Co. Limited, Toronto.

Library of Congress Cataloging in Publication Data

Morris, Richard, date.
Evolution and human nature.

Includes index.
1. Sociobiology. 2. Human evolution. I. Title.
II. Title: Human nature.
GN365.9.M67 1982 304.5 82-19572
ISBN 0-399-31010-X

Evolution and Human Nature

Preface

I suppose I ought to begin by explaining why I wanted to write a book about evolution and human nature. After all, I was trained as a physicist, not as an anthropologist or a biologist.

Obviously the topic interests me. In fact, I have always found the subject of human nature to be fascinating. When I was still an adolescent, I immersed myself in books on Freudian psychology. At the time, I thought that psychoanalytic theory was the only kind of psychology there was. I knew nothing, for example, about the behaviorist psychology. I wasn't yet aware of the fields of sociology and anthropology, which also deal with the behavior of human beings. I didn't even realize that the great novelists and poets could tell me as much about human nature as the behavioral scientists.

Nearly thirty years have passed since I began reading popular accounts of Freud's theories. More than twenty-five have gone by since I stopped believing in them dogmatically. In all that time, my interest in human nature has not diminished. I still find myself drawn to books that promise to give me insight into the workings of the human mind. I find myself reading them as voraciously as I ever did. The only difference is that I no longer believe that everything can be explained in terms of Oedipus complex, id, ego and superego.

But this isn't a book about Freud. In fact, his name crops up only in a few scattered places. Nor is it a book about any of the various kinds of psychology that are practiced in the universities.

Academic psychology has not yet unraveled the intricacies of human motivation. I do not mean to imply that psychology has not had its successes. It has delved quite deeply into such topics as the conditioning of experimental animals and into human perception, to give just two examples. However, it has done little to explain the biological influences on human behavior.

And what does biology have to do with human nature? Simply this: if there is any such thing as a basic human nature that is not the result of cultural conditioning, it is something that has evolved by natural selection. Since the time of Darwin it has been recognized that evolution molds the behavior of animals; it does not act on physical structure alone. Not only does Darwin's theory of natural selection tell us why a woodpecker has a certain kind of beak and why hawks have keen eyesight, it also explains why it is that baboons go about in troops and why seals fight one another for possession of breeding territories. As bodies evolve, so do brains. And brains have certain innate predispositions and behavior patterns programmed into them. A bird, for example, does not have to learn how to build a nest; evolution has stamped that knowledge into its brain.

There is no human behavior that is analogous to nest-building in birds. Our behavior is more flexible, more dependent upon learning. Since our large brains make it possible for us to learn so much, we have less need for behavior patterns that are "wired in."

Human beings are different from all other animals in yet another way. Only humans have culture. This also implies that genetically controlled behavioral patterns and predispositions should play a less important role in humans than they do in other species. It is culture, not physical evolution, that makes it possible for us to adapt ourselves to so many different kinds of environment. And it is culture—not our genetic endowment—that teaches us to accept certain values and goals.

Nevertheless, there are certain kinds of human behavior that are innate. We all know how to laugh, for example. Even the congenitally deaf, who have never heard laughter, can do that. We all feel anger. We all know the meaning of grief. Except for a few individuals who are afflicted by certain kinds of mental ailments, we are all capable of compassion. It is true that learning has some effect on the expression of these emotions; our cultures, for example, teach us to be amused by certain things. But they do not tell us what laughter is. That is something that we know long before we learn to speak.

But some questions about human nature are not so easily answered. For example, what is human aggression? Is it learned? Is it innate? Do we have an aggressive drive that must, one way or another, find an outlet? And what about love? Is the capacity to love programmed into us? Are there fundamental emotional or intellectual differences between men and women? Are there genes for human altruism? Are there genes that create predispositions toward antisocial behavior? Are human beings naturally monogamous?

Questions like these have been the object of considerable controversy for more than a century. This book is an attempt to explain what these controversies are all about. Although the chapters follow one another in a kind of historical progression, it was not my intention to write a comprehensive intellectual history. I simply felt that one must know something about the arguments that raged in the past before one can properly understand some of the controversies of the present.

This is not a book about politics or about scientific mistakes, although both topics are discussed at some length. Political questions came up because theories about what a human being is lead naturally to theories about the manner in which human society should be structured. Theories of human nature have been used to justify democracy, socialism, fascism, colonialism, capitalism and racism. They have been used in arguments about the necessity for revolution, and they have been used to bolster conservative apologies for the status quo. It would be possible to discuss human nature without making any mention of politics, but that would be like describing the ecology of a forest without making any mention of the fact that forests are made up of trees.

My reasons for bringing in political ideas should be fairly obvious. My emphasis on scientific mistakes requires, perhaps, a little more justification. Few books dwell on misconceptions that have been created by inaccurate data or on mistaken theories. After all, we are generally more interested in the successes of science than we are in all the blind alleys that have to be blundered into before significant discoveries are made. Dwelling on errors is ordinarily a poor way to expound on truths.

Nevertheless, this is exactly what I have done. In fact, it is sometimes scientific errors and distortions that have received the greatest emphasis. There are a number of reasons for this. First, some incorrect theories of human nature have been quite influential. In some cases, their influence persists today. Second, many of us—scientists included—have an emotional investment in our ideas about human nature. As a result, prejudices and predispositions have played a more important role in theorizing about human nature than they have in other scientific endeavors.

No scientist is ever completely objective. Since he is human, he cannot avoid being influenced by preconceptions, desires, intuitions and feelings.* In fact, preconceptions play an important role in scientific discovery. Without them, hypotheses would never be formed. Only when one has intuitive feelings about what *ought* to be true does it become possible to construct a theory.

Naturally, many theories turn out to be wrong. But even incorrect theories play an important role. They tell scientists what experiments should be performed and what observations should be made. These experiments and observations often point the way to ideas that are more nearly correct than those they supplant. Science could not proceed in any other way.

Preconceptions have played an especially prominent role when scientists have turned their gaze upon themselves. This is only to be expected. If preconceptions about how an atom should behave

*Throughout this book I use the generalized masculine pronoun to mean both male and female individuals. Although phrases such as "he (or she)" have the advantage of being non-sexist, they sound awkward. I hope that my women readers will not consider this a slight. It is not intended to be one.

can sometimes lead a physicist astray, then the scientist who studies human nature must be especially susceptible to being misled.

In particular, Victorian theories about evolution and its influence on human nature reflected Victorian prejudices. Liberal political outlooks have led to liberal interpretations of scientific data. In years gone by, when Western culture was racist, scientific theories too were racist. There is nothing surprising about this. While we are still children, our culture teaches us to view human nature in certain ways. Those children who eventually become scientists are subjected to the same cultural conditioning as is given to those who go into other occupations.

But this book was not written as an exposé of scientific errors. Like everyone else, scientists learn from their mistakes. As a result, down through the years they have been gradually blundering their way toward the truth. It is possible to argue that they still do not know what human nature is. However, one cannot deny that they have learned quite a bit about what it isn't. As we shall see, that is quite an accomplishment.

This brings me to my other reason for writing this book. I like to think that having a scientific background outside the biological and behavioral sciences has made it possible for me to write about such things with a reasonably clear perspective. I undoubtedly have preconceptions of my own. However, I have not been a participant in any of the current controversies about human nature, and I have not done scientific work that caused me to become committed to a certain kind of view. I cannot view theories of human nature as objectively as the proverbial man from Mars, but the fact that I was trained as a physicist, not as a biologist, does make it easier to maintain a certain distance from the subject.

At least that is what I tell myself. I imagine that, in the end, it will be my readers and critics who must decide whether or not that is really true.

1

Two Theories of Human Nature

The word *philosopher* conjures up the image of a person who lives a scholarly life. Today, when almost all philosophers are university professors, the word makes us think of someone who spends a significant part of his life in libraries, who revels in making hair-splitting distinctions, who inhabits a world of the intellect.

If this picture is something of a stereotype today, it was even less accurate during the seventeenth century. For example, the life of the English philosopher John Locke was anything but scholarly. Locke was so caught up in the political intrigue of the day that he published nothing until he was fifty-six years old. He was forced to spend six years in exile in Holland. During that time, he met with conspirators against the English Crown. At one point, when the English government was trying to have him extradited so that he could be tried on a charge of treason, he went into hiding.

From 1667 to 1683, Locke was adviser and personal physician to the Earl of Shaftesbury, leader of the Whig party in Parliament. At the time, Parliament and King Charles II were engaged in a bitter struggle for political power. Charles was trying to establish the principle that kings rule by divine right, while Parliament struggled to assert its own supremacy.

The seventeenth century was a turbulent era in England. The politicians of the day engaged in plots and intrigue. When they could not attain their ends by conventional means, they did not hesitate to resort to conspiracy. Serving in Parliament sometimes led to imprisonment in those days. Shaftesbury himself was confined in the Tower of London from 1677 to 1678. Upon his release, his fortunes changed, and Charles appointed him President of the Privy Council. But he was arrested again in 1681 and charged with treason. Although he was later released, Shaftesbury found it wise to flee England early in 1683 and seek refuge in Holland. He was in Holland when, later the same year, he died.

Shaftesbury was involved in plots to place Charles's illegitimate son, the Duke of Monmouth, upon the English throne. He and his co-conspirators wished to prevent the succession of Charles's heir, the Roman Catholic Duke of York. This conspiring led eventually to the Rye House Plot, an attempt to assassinate York and the King in 1683, and later to an armed rebellion upon the death of the King in 1685.

It cannot be proved that Locke was involved in any of these conspiracies. However, there is a certain amount of circumstantial evidence indicating that he probably was. Shaftesbury appointed Locke to various government posts, solicited his advice on political matters, and sent him on errands of a political nature. Shortly after Shaftesbury fled England, Locke found it wise to do the same. Before he left, he was careful to destroy a quantity of papers.

When Charles died in 1685, the Duke of York succeeded to the throne as James II. But he was to rule for only three years. James was deposed in the Glorious Revolution of 1688, after which Parliament offered the throne to William of Orange and his wife, Mary. They accepted, and ruled jointly as William III and Mary II. Under the new regime, Parliament's powers were increased. The Glorious Revolution can be said to mark the beginning of constitutional monarchy in England.

Locke returned to England in 1688. Within a short time, he had established himself as the philosopher of the revolution. In 1690 he published his *Two Treatises of Civil Government*. The *Treatises* were attempts to justify the 1688 revolution and the supremacy of Parliament.

Since it is Locke's theory of human nature, not his political philosophizing, that interests us, I will not describe the contents of the *Treatises* in detail. It is worth mentioning, however, that the second treatise, a theoretical justification of parliamentary government, was quite influential. Some of the principles that Locke enunciated found their way into the American Declaration of Independence and later into the United States Constitution. They were also widely discussed by the makers of the French Revolution. Finally, Locke's doctrine of checks and balances is still very influential today.

In the same year that the *Treatises* appeared, Locke published his most important work, *An Essay Concerning Human Understanding*. This massive work, which is an exposition of Locke's theory of human nature, had an enormous impact on political thought. Paradoxically, it contained nothing about politics at all.

It was Locke's intention to determine the nature and limits of human knowledge. He began by rejecting the idea, which was commonly held in his day, that anything could be known innately. On the contrary, Locke said, all human knowledge came from experience. The human mind, he maintained, was a piece of "white paper" (or, as we might say today, a "blank sheet") upon which sense impressions were inscribed. Everything that a human being could think of—even very abstract conceptions—was the result of perceptions and the operations of the human mind. According to Locke, the senses provided us with simple ideas, such as *yellow*, *white*, *heat*, *cold*, *soft*, *hard*, *bitter* and *sweet*. The mind operated upon these simple concepts to form more complicated ones. It was experience that provided the foundation for all human thought.

Although the idea that the human mind was a *tabula rasa* (Latin for "blank slate"; its meaning is the same as Locke's "white paper") at birth was not original with Locke, it was he who worked out the theory in detail. One would not think that this empirical approach to knowledge would have profound implications for politics. Locke's contemporaries thought it did, however. It was the white-paper theory that led to the notion of *equality* that played so important a role in the American and French revolutions.

Until Locke's time, it had generally been believed that there were innate differences between the members of the aristocracy and the remainder of the population. Aristotle had maintained that the noble and wise had an inherent right to govern. During the next two millennia, few had questioned that view. And then, suddenly, it was upset by a philosophical theory of human nature, a theory that said that everything we know is the product of experience, that it is learning, not heredity, that makes us what we are.

If Locke's white-paper theory was correct, his contemporaries reasoned, then there was no such thing as hereditary nobility or virtue or wisdom. If environment, not birth, was responsible for the differences between human beings, then no particular class could claim the right to rule. The members of the aristocracy did not have superior "blood," only a better education.

"Can there be anything more splendid," asked the French philosopher Voltaire, "than to put the whole world into commotion by a few arguments?" Voltaire was not exaggerating. Locke's philosophy did put the world into commotion. If the mind was originally blank, then it seemed to follow that men (women were often left out of the argument) were inherently equal. If education and environment made human beings what they were, then it should be possible to create an ideal society. If one created the right social influences, then that was all that would be needed to make men perfect.

The view that Locke worked out in his *Essay* was only a theory. But it had a revolutionary impact. The implications of Locke's philosophy dominated eighteenth-century thought, and they continue to be influential today. The French revolutionary slogan "liberty, equality, fraternity" can be traced back to Locke, as can the statement in the American Declaration of Independence that "all men are created equal." The modern idea that society's ills can be cured by social legislation that provides individuals with better environments is a consequence of this interpretation also.

Locke's ideas were somewhat misinterpreted by his followers. In particular, he never said that all men were equal. On the contrary, he believed that there were differences in human capacity and temperament. Furthermore, he did not believe that governments should treat all people equally. Locke thought that women and the poor should be excluded from participation in politics; in his view, only male property owners should be granted the right to vote.

Locke's disciples found the idea that human qualities were the result of environment to be so intoxicating that they ignored the more conservative elements in Locke's thought. The new ideas about the nature of the human mind had such a liberating influence upon them that they did not often trouble themselves to familiarize themselves with what the English philosopher had actually said (this shouldn't be considered especially surprising; today there are numerous Marxists who have never read Marx).

In his own day, Locke frequently was compared to Newton. The two Englishmen were classed together by those who would believe that one had discovered the laws of nature, while the other had found the laws which governed the human mind. One had penetrated the mysteries of gravitation; the other had shown that society could be reconstructed on rational lines.

As we are all aware, society was not reconstructed in the years that followed. There were political transformations. The rising middle class was successful in seizing power from the hereditary aristocracy. But the ideal of perfect social equality was, except for a few brief periods, never put into practice. When the Industrial Revolution arrived, new masters were substituted for the old. Farm laborers left the countryside to find work in the cities, where they were forced to work twelve- and fourteen-hour days in factories. They were not allowed either to form unions or to vote. Even in the United States, where so much lip service was paid to the idea of "equality," the founding fathers established property qualifications for voting.

By the time the Victorian age arrived, it seemed that the revolutionary era had passed. The pendulum had begun to swing the other way. The idea that heredity played an important role in the formation of human character again became fashionable. The age demanded a theory of human nature that justified the status quo.

The most influential proponent of a hereditarian view of human nature was Sir Francis Galton. Galton, who was a cousin of Charles Darwin, had perhaps the most diverse interests of all the Victorian scientists. By the age of twenty-two he had completed his medical studies and had earned a mathematics degree from Trinity College, Cambridge. By the time he was thirty, he had become known as an African explorer. During the remainder of his long life, he made contributions to the fields of meteorology and statistics, helped to establish the importance of fingerprints as a method of identification, and became one of the first English experimental psychologists.

Today Galton is best remembered for his studies of heredity and for his founding of the eugenics movement. Eugenics was, as one of Galton's followers put it, "the science of the improvement of the human race by better breeding." Believing that heredity was much more important than environment, Galton concluded that efforts should be made to improve humanity's genetic makeup. If this was not done, he thought, then "racial degeneration" was inevitable.

Today such ideas seem questionable and are tinged with racism as well. But in Galton's time, many found them to be perfectly reasonable. During the Victorian age, very little was known about the mechanisms of genetic inheritance, and it seemed natural to believe that it should be possible to create a better society by encouraging people with desirable traits to have more children and discouraging those who had inherited inferior qualities.

Selective breeding was first advocated in Galton's influential book *Hereditary Genius*, which was published in 1869. In this work, Galton attempted to show that exceptional ability ran in families. The offspring of eminent men, he claimed, had a better-than-average chance of achieving eminence themselves. In order to determine that this was indeed the case, Galton examined the achievements of some thousand individuals in three hundred different families. Some of these people were distinguished judges or statesmen. Others were noted novelists, military commanders, scientists, poets, musicians, painters and clergymen. Finally, just to make his study complete, Galton considered the inheritance of ability in oarsmen and wrestlers.

Galton found that men of exceptional ability were often related. Although the correlation was strongest in the case of fathers, brothers and sons, more distant relations of eminent people also tended to be very successful. Some 31 percent of the fathers of eminent men had achieved eminence themselves, as did 41 percent of the brothers and 48 percent of the sons. When one considered nephews, the figure fell to 22 percent, and it was only 13 percent for first cousins. But this only seemed to confirm the hypothesis. It stood to reason, Galton thought, that the more distant the relationship, the more the inherited ability would be diluted.

According to Galton, intelligence was not the only quality that was inherited. He believed that his results also showed that such things as piousness and an interest in science were hereditary. Furthermore, judges tended to father judges. Statesmen begat statesmen, and literary ability in a father often produced similar qualities in his offspring. Naturally, there were exceptions. Some of the relatives of judges distinguished themselves as bishops, admirals, poets and physicians, for example. However, the statistical patterns seemed to be clear.

Today we do not find it surprising that a certain number of people follow in their fathers' footsteps. And if we observe that the offspring of eminent men tend to be successful also, we hesitate to conclude that this fact can be ascribed wholly to heredity. There is such a thing as an advantageous environment, after all. And of course, a successful parent is likely to see that his children receive a better-than-average education. If a study like Galton's were performed today, we would tend to think that it proved very little. We certainly would not believe that it showed that people who are exceptionally successful achieve their positions through superior heredity alone.

Yet this is exactly the conclusion that Galton reached. He made the assumption that the people he studied would have become eminent even if they had had disadvantageous environments. In Galton's view, genius was something that allowed one to overcome all obstacles. If one was born with superior character and intelligence, he would eventually find his way to the top, whether he came from a long line of earls or was the son of a laborer. Galton had little patience with the view that all babies were born pretty much alike. Nor did he believe that environment or education could develop qualities that were not already present in an individual. The only thing that mattered was innate ability. Furthermore, eminence was a criterion by which ability could be judged.

Such views exhibit a certain amount of class bias. If innate ability is the only thing that matters, then it follows that the lower social classes must be made up of individuals who are less capable than those who occupy the more privileged positions in society. In Galton's view, the fact that the working classes produced few eminent people was evidence of their inferiority. If social obstacles really did prevent people from rising to the top, he argued, then there would be more eminent men in America, where the barriers to advancement were fewer, than there were in England. Social advantages, he maintained, did not give eminence to those of moderate ability.

Galton divided human beings into sixteen different classes according to their intelligence. Those who were above average were placed in classes labeled with the letters A through G, and X. The categories of below-average intelligence were designated by the corresponding lower-case letters. According to Galton, about one person in four thousand fell into one of the three highest classes, F, G and X. The X category was occupied by one in a million. Idiots and imbeciles, on the other hand, could be designated by the letters f, g and x.

Galton thought that some dogs were more intelligent than the lowest-ranking humans. "I presume the class F of dogs and others of the more intelligent sort of animals," he said, "is nearly commensurate with the f of the human race, in respect to memory and powers of reason. Certainly the class G of animals is far superior to the g of humankind."

Galton also believed that some black people were far more intelligent than the average white person. But he was not willing to grant blacks equal intellectual status. On the contrary, he thought that, on the average, they ranked two classes below whites. On the other hand, he placed the white Englishmen of his day two full classes below the inhabitants of ancient Athens.

Galton did not have a high opinion of those who made up the middle categories, b, a, A and B. Their intellectual grasp was "feeble and hesitating," he judged, and they were not well adapted to the complexities of modern civilization. If civilization was not to decay, a way must be found to upgrade them.

Galton believed that this could be done by instituting selective breeding. The genetically fit, he thought, should be encouraged to marry early and to have large numbers of children. The inferior, on the other hand, should be discouraged from producing offspring. If they continued to "procreate children, inferior in moral, intellectual and physical qualities," then they should be considered to be "enemies to the State." Presumably they would be dealt with accordingly.

Galton thought that he could foresee the day when a new kind of caste system would be set up. Registers of the "hereditarily remarkable" would be drawn up, and certificates would be awarded to the gifted. As the political influence of the elite class increased, it would be able to enact legislation beneficial to its members. Laws would be passed to prohibit the nongifted from inheriting fortunes, and endowments would be set up to enable the elite to multiply more rapidly. The elite would, of course, continue to treat their inferiors "with all kindness, so long as they maintained celibacy."

If eugenic programs were conscientiously carried out, Galton admitted, then the "inferior" races would gradually be replaced. He did not think that this would be any great loss, although he did realize that such a program might encounter resistance. Galton argued for such a program of genocide in his book *Inquiries into Human Faculty and Development*. His comments on the subject are worth noting at length:

> There exists a sentiment, for the most part quite unreasonable, against the gradual extinction of an inferior race. It rests on some confusion between the race and

> the individual, as if the destruction of a race was equivalent to the destruction of a large number of men. It is nothing of the kind when the process of extinction works silently and slowly. . . . That the members of an inferior class should dislike being elbowed out of the way is another matter; but it may be somewhat brutally argued that whenever two individuals struggle for a single place, one must yield, and that there will be no more unhappiness on the whole, if the inferior yield to the superior than conversely, whereas the world will be permanently enriched by the success of the superior.

Galton, who was an agnostic, believed that eugenics might eventually take the place of religion. "An enthusiasm to improve the race is so noble in its aim," he wrote, "that it might well give rise to a sense of religious obligation." Galton certainly acted as though eugenics was a religion. During the last decade of his life, he proselytized endlessly, arguing his ideas in books, in essays, and in addresses to gatherings of scientists.

Although Galton's ideas were by no means universally accepted, they were taken seriously. When the Eugenics Education Society was founded in 1907, a number of eminent scientists became members, as did the novelist H. G. Wells, the playwright George Bernard Shaw, the socialist Sidney Webb, and numerous members of the nobility.

But it was in the United States that the eugenics movement was most influential. Unlike some of their European counterparts, the American eugenists did not confine themselves to promoting their doctrines in essays and in books, they also lobbied for the passage of sterilization laws and were influential in persuading Congress to place restrictions on immigration during the 1920s.

The American eugenists were especially concerned about what they called the "menace of the feeble-minded." They believed that such social ills as crime, alcholism and prostitution were related to subnormal intelligence, and advocated sterilization to prevent the unfit from breeding.

By 1917, sterilization laws had been enacted in some sixteen states. All these laws provided for the sterilization of the mentally retarded, and most applied also to epileptics and the "insane." Other classes of people that were affected by the various laws included confirmed criminals, alcoholics, syphilitics, "moral degenerates" and prostitutes. One bill, introduced by a Missouri legislator, would have provided for the sterilization of anyone "convicted of murder (not in the heat of passion), rape, highway robbery, chicken stealing, bombing or theft of automobiles."

Some 12,145 sterilizations were carried out between 1907, when the first law was passed, and 1931. But the number was much lower than it might have been. In some states the laws were not enforced. In others, sterilization statutes were held to be unconstitutional by state or federal courts.

Doubts about the constitutionality of sterilization were banished, however, when the United States Supreme Court upheld the Virginia sterilization law in 1927. The Court's opinion was delivered by Justice Oliver Wendell Holmes, who wrote:

> We have seen more than once that the public welfare may call upon the best citizens for their lives. It would be strange if it could not call upon those who already sap the strength of the State for lesser sacrifices. . . . It is better for all the world, if instead of waiting for their imbecility, society can prevent those who are manifestly unfit from continuing their kind.

By 1958, sterilizations had been ordered for more than sixty thousand people. Between 1931 and 1958, 48,781 operations were performed—12,463 in California alone. The Virginia law that Holmes upheld was enforced until 1972. More than seven thousand were sterilized in that state; many of those subjected to the operation were not even told what was being done.

But it was in the area of legislation on immigration that the eugenics movement had the greatest influence. During the years before World War I, scientists who participated in the movement became concerned about the supposed inferior intellectual qualities of recent immigrants. According to the psychologist Henry H. Goddard, one of the eugenists' spokesmen, as many as 50 percent of those entering the United States were feeble-minded.

This view was elaborated upon by such eminent men as Princeton University psychologist C. C. Brigham and by Edward A. Ross, a sociologist at the University of Wisconsin. Making use of racial theories that were commonly believed at the time, these men argued that all European peoples were mixtures of three races, the Nordics, the Alpines and the Mediterraneans. The Nordics, who came from northern Europe, were a race of "rulers, organizers and aristocrats"; they were much superior to the Alpine peasants. The short, dark Mediterraneans were the worst of the lot; addicted to crimes of sex and violence, they tended to have "low foreheads, open mouths, weak chins, poor features, skew faces, small or knobby crania, and backless heads."

Changes in immigration patterns, the eugenists feared, were causing America to be inundated with inferior specimens of humanity. If immigrants from central and southern Europe continued to stream into the country, then the old Nordic stock would find that its purity had been tarnished. Intermarriage and "mongrelization" would result, and the quality of American civilization would inevitably decline.

Although some prominent eugenists dissociated themselves from such racist views, sentiment in favor of limiting immigration continued to grow. During the 1920s Congress was persuaded to pass a series of laws that placed limits on immigration and set up quotas for individuals of different national origins. The first act, passed in 1921, instituted quotas. In 1924 the quotas were reduced. Limits were set upon immigration from central and southern Europe, and Orientals were barred completely. Although difficulties in estimating the national origins of the American people prevented the new policy from being enforced immediately (the quotas were supposed to be related to the number of people of specified national origins already in the country), estimates eventually were made, and Congressional approval was obtained. The enforcement of national-origin quotas became official American immigration policy on July 1, 1929.

During the 1930s interest in eugenics declined rapidly. Developments in such fields as anthropology and genetics demonstrated that eugenics did not have a firm scientific foundation. On one hand the anthropologists showed that environmental and cultural factors played a greater role in forming human character than the eugenists had thought. Research in genetics demonstrated that human heredity was much more complicated than had been supposed. In particular, "feeble-mindedness" was not a trait that was inevitably passed on from parents to offspring, while superior qualities were the result of a complex interplay of heredity and environment. Finally, the Nazis in Germany showed how eugenic doctrines could be carried to extremes. By the end of the decade, the movement had withered away. However, the sterilization and immigration laws remained.

Today it is apparent that eugenics was based upon an assumption that was rather questionable, to say the least. The questionable assumption was not that differences between individuals reflect genetic endowments—nothing could be more obvious than that fact—but that inherited differences are not much affected by environment.

Galton thought that individuals with exceptional inherited ability would inevitably rise to positions of eminence, while the lower rungs of society would be occupied by those who were innately inferior. He believed that intellectual and moral qualities alike were determined by heredity, and that there was such a thing as an inherited criminal nature. His disciples never questioned these ideas. On the contrary, they elaborated upon them. And when they were confronted with evidence that seemed to show that environmental factors would be important, they attempted to explain that evidence away.

For example, when it was observed that immigrants who had lived in the United States for a number of years scored higher on I.Q. tests than those who had only recently arrived, the eugenists did not draw the natural conclusion that familiarity with American culture influenced scores. They concluded, instead, that the quality of the immigrants was declining. Although eugenics never became a religion as Galton had hoped, its tenets certainly took on the character of dogma. When the evidence seemed to contradict them, it was the evidence, not the eugenic doctrine, that was reinterpreted.

But Locke's white-paper theory too has occasionally become a dogma. In the eyes of some of the eighteenth-century political theorists, it provided a justification for revolution. It is hard to imagine any evidence that might have caused them to give up the theory. Revolutionaries are not that objective about their beliefs. Today, Locke's theory is often an integral part of the thought of social reformers who believe that social ills can be cured by improving people's environment. They would not readily give up this idea either.

The point that I am trying to make is not that one of these two theories of human nature is correct and the other wrong. Since the late 1940s, it has been apparent that human character and intelligence are not molded primarily by heredity, or primarily by environment, but by a complex interaction between the two. In most cases it is not possible to ascribe human traits to one or to the other.

Many people find it natural to believe that intelligence is hereditary. However, it has been shown that scores on I.Q. tests (which may or may not measure the quality that we call "intelligence") are heavily influenced by environmental factors. On the other hand, personality is commonly thought to be the result of environmental influences. But there are studies that seem to indicate that infants differ in temperament from birth. Heredity and environment can rarely be separated. As a result, many scientists have long maintained that the so-called nature-nurture controversy is dead.

If it is dead, it is a very lively corpse. As we will see in subsequent chapters, arguments about the relative importance of environmental and genetic factors are still being made. They constitute an important element in the current controversy over sociobiology, for example. It appears that the nature-nurture controversy has not yet been settled.

The arguments are all the more vehement because political commitments enter in. Theories about human nature do have implications for politics. It is no accident that Locke's ideas were taken up by the radicals who made the French and American revolutions, or that Galton's theories should have appealed to political conservatives and to those who sought scientific justification for their ethnic and racial prejudices.

Scientists have been swayed by their political inclinations as frequently as has anyone else. At least this was the conclusion reached by the American psychologist Nicholas Pastore in his book *The Nature-Nurture Controversy*, which was published in 1949. Pastore examined the theories and political leanings of some twenty-four English and American scientists who had aligned themselves on one side or the other. He found that, with one exception, those who thought that heredity was more important were political conservatives. On the other hand—again with one exception—the environmentalists were all liberals or radicals.

Pastore realized that he had demonstrated only that there was a correlation. His data did not allow him to state with any degree of certainty either that political leanings had influenced scientific outlooks, or that the reverse was the case. Nevertheless, he did state an opinion. In his view, "the sociopolitical allegiances of the scientists were a significant determinant of their position on the nature-nurture question." Furthermore, "these allegiances had a marked effect upon the formulation of a hypothesis and the method of its verification." In other words, these scientists were sometimes influenced less by the intrinsic scientific merits of the various theories that were put forward than they were by their political loyalties.

It certainly seems to be apparent that—at least in the cases of Locke and Galton—political leanings played an important role in the formulation of supposedly objective scientific theories. It would be unreasonable to assume that Locke's liberal political views had nothing to do with his conception of human nature. Galton, on the other hand, was a well-to-do Englishman who had a conservative outlook and a low opinion of the lower classes in Victorian England. It is not surprising that he should have come up with a theory that equated differences in social class with differences in inherited character. Intelligence was not the only quality that Galton thought to be inherited. He also included such qualities as "civic prosperity" and "civic worth."

And what, exactly, is human nature? It may seem that this question has hardly been touched upon. So far, we have only considered the question of whether it is something that is learned or something that is inherited, and we haven't even found the answer to that. Instead, we have found ourselves embroiled in discussions of political issues. We have been able to observe very little in the way of scientific fact. We have found only that scientists and lay people alike have been influenced by their prejudices, preconceptions and political beliefs. We have found that theories have contained biases, and that facts have been misinterpreted to fit preconceived notions.

Admittedly, human nature is an explosive subject. Prejudices, dogmatic beliefs and emotional commitments to certain kinds of outlook play a role even today. However, we ought to be able to find out something about ourselves by delving into the subject of human evolution. After all, if any part of human nature is hereditary, it must be something that has evolved.

Discussions of evolutionary theories of human nature will occupy the greater part of the remainder of this book. However, we should not deceive ourselves into thinking that the nature-nurture controversy has been discussed and disposed of. As we shall see, it is something that has cropped up again and again.

❦ 2

Theories of Evolution

Whenever we think of evolution, the name of Charles Darwin comes to mind. Yet Darwin did not invent the idea. When he published his book *On the Origin of Species* in 1859, philosophers and scientists had been speculating about the evolution of life for more than two thousand years.

Evolutionary ideas made an appearance in ancient Indian and Chinese thought and in the teachings of the pre-Socratic Greek philosophers. Although the opposition of Plato and Aristotle temporarily brought evolutionary speculation to a halt, at least in the Western world, the idea had resurfaced long before Darwin wrote his book. Around the middle of the eighteenth century, the French naturalist the Comte de Buffon suggested that animals might be altered by changes in the environment. He even hinted that man and the apes might have a common ancestor.

But Buffon did not weave his ideas into a consistent theory. Perhaps he feared the opposition of the Church, or perhaps he was not very sure about the validity of his speculations. In any case, his evolutionary ideas are scattered throughout the forty-four volumes of his *Natural History*. What is worse, he sometimes contradicts himself, denying what he has gone to so much trouble to suggest.

Even if Buffon had brought his ideas together in a single, well-reasoned volume, it is unlikely that they would have been accepted. Although Buffon speculated about evolutionary changes, he had no theory that explained how these changes took place.

Dr. Erasmus Darwin, Charles's grandfather, was also an evolutionist. In his *Zoonomia,* published in 1794, he proposed that "irritations, sensations, volitions and associations" caused animals to adapt themselves, and that the resulting changes were passed along to offspring. Dr. Darwin estimated that the earth was "millions of ages" old, and suggested that all life might have descended from a single organism.

A somewhat more elaborate theory of evolution was proposed some fifteen years later by the French naturalist and philosopher Jean Baptiste de Lamarck. Unlike Dr. Darwin, Lamarck did not believe that all living creatures had a common ancestor. On the contrary, he thought that life would arise from nonliving matter by a process known as spontaneous generation. In Lamarck's theory, life had arisen, not once, but many times.

According to Lamarck, all creatures had a built-in drive toward perfection that caused them to rise on the evolutionary scale. The lower forms were continually transforming themselves into higher ones. Meanwhile, spontaneous generation filled the resulting gaps at the bottom of the scale. If the modern conception of evolution can be likened to a tree in which the different branches represent various evolving species, then Lamarck's conception can be compared to a set of parallel ladders. In his theory, the individuals who made up the various living species had different ancestors, and these had begun to evolve at different times.

Today Lamarck is often associated with the discredited idea that evolution can proceed by means of the inheritance of acquired characteristics. Although this was indeed part of Lamarck's theory, the idea was not original with him. When Lamarck incorporated the concept of inheritance of acquired characteristics into his theory, he was only adopting a notion that was very much a part of eighteenth-century thought.

According to this theory, the reason that giraffes have long necks is that their ancestors were always stretching theirs in order to eat foliage from trees. Similarly, a man or woman who spent a lot of time lifting heavy weights would be likely to have unusually strong offspring. Today we know that this idea is false. The use or disuse of certain parts of the body cannot affect the *gametes*, or sex cells, that are formed in the testes or ovaries.

Charles Darwin did propose that the cause of evolution was a mechanism that he called *natural selection*. But even this was not original with Darwin. As William Irvine remarks in his book *Apes, Angels and Victorians*, the possibility of evolution by natural selection was something that had occurred to half the crackpots in Europe by the time Darwin's theory was published. For example, the theory of evolution by natural selection was clearly stated by an English author named Patrick Matthew in the appendix of his book *Naval Timber and Aboriculture*, which appeared thirty years before Darwin's *Origin of Species*.

Although Matthew went so far as to have "Discoverer of the Principle of Natural Selection" printed on his visiting cards, no one paid very much attention to his claim at the time, and no one does so today. The credit for discovering natural selection is rightly given to Darwin, because Darwin did much more than make a few offhand suggestions. He spent decades amassing evidence that would convince the scientific world that both evolution and the natural-selection theory were plausible ideas.

Evolution and natural selection are not the same thing. The former is the transformation of species over the course of many generations. The latter is the two-part process that causes these changes to take place. Although we often speak of "the theory of evolution," we are being somewhat careless when we do. Evolution is a fact; natural selection was the theory that was invented to explain it.

According to Darwin's theory, evolution is the result of a struggle for survival. Darwin noted that the amount of variation in all known species was enormous. All members of a given species were not the same; individuals differed from one another in numerous ways. Darwin also observed that members of all species produced more offspring than could possibly survive. Only some of the offspring grew to maturity and reproduced. If the differences between individuals were inheritable, Darwin reasoned, then it followed that certain advantageous hereditary characteristics would be passed on to subsequent generations more often than others. The best-adapted individuals would have the largest number of progeny. Therefore the characteristics possessed by the best-adapted would be most likely to appear in successive generations. This would ensure that, over long periods of time, species would evolve.

Note that two things are required if there is to be any natural selection. First, there must be inheritable variation. Second, there must be selection. Darwin's great insight was the realization that random variation had to come first. Differences between organisms had to exist before there could be any selection. Darwin's theory was thus quite a bit different from that of Lamarck. According to the latter, an upward evolutionary striving came first. Variations were the result.

Natural selection is often summed up by the phrase "survival of the fittest." Although Darwin used this term, he did not coin it. It was invented by a contemporary, the English philosopher Herbert Spencer, who used it before Darwin's *Origin of Species* was published.

"Survival of the fittest" is not a very apt description of natural selection. In fact, if one is not careful, it can lead to circular arguments. Selection ensures that the "fittest" survive? But who are the "fittest"? They are the animals with the best potential for survival. "Survival of the fittest" seems to mean nothing more than "survival of those best able to survive."

Fortunately, there is a much better way to define evolutionary fitness. A modern biologist would say that the fitness of an animal is measured by the number of viable offspring it produces. If an individual does not have offspring, it makes no difference whether it dies in infancy or lives to a ripe old age. In either case, it will not have passed any inheritable characteristics on to the next generation. The distinction was not always made in Darwin's day, and it did lead to some misinterpretations. However, the circularity of the term (which was not even used in the first edition of *Origin of Species*) does not imply that there are any defects in Darwin's theory.

Although Darwin thought of the idea of natural selection in 1838, twenty-one years passed before his theory was published. He did write a 35-page sketch of the theory of natural selection in 1842, and a 231-page essay in 1844, but neither of these found its way into print. In 1846, after giving his wife instructions to the effect that the 1844 essay should be published in the event of his death, Darwin put evolution aside and embarked on a study of barnacles that was to consume the next eight years of his life. Only after the work on barnacles was published in four volumes between 1851 and 1854 did Darwin return to the subject of evolution.

In 1854, Darwin began writing a lengthy treatise on the subject. But the work did not go quickly. Several more years went by. The book grew longer and longer, but it was nowhere near completion. And then, around the end of 1858, Darwin's colleague Alfred Russel Wallace wrote a letter that was accompanied by a paper on the subject of evolution by natural selection. Wallace asked Darwin to look the paper over, and to send it on for publication if it seemed worthy.

Although Darwin had formulated the idea of natural selection first, Wallace had developed a similar theory independently and had been the first to prepare his ideas for publication. Darwin and Wallace could easily have entered into a bitter fight over priority. Such battles are not uncommon in science.

Fortunately, neither felt inclined to pursue such a course. At first, Darwin felt that the only honorable course was to refrain from publishing his own theory. When he changed his mind and allowed his friends the geologist Charles Lyell and the botanist Joseph Hooker to present a sketch of his theory to the Royal Society along with Wallace's paper, Wallace graciously declared that this action had been more than generous. Rather than press his own claims, he acknowledged Darwin as the true discoverer of the concept of natural selection.

Shortly after the joint presentation was made, Darwin began to write a brief paper of his own. Like the treatise that he had been writing earlier, it grew longer and longer. But Darwin was able to finish it this time. On November 24, 1859, the "brief sketch" had grown into a sizable book and was published under the title *On the Origin of Species*.

Numerous scholars have speculated about the reasons for Darwin's twenty-one-year delay in bringing his ideas before the world. It has been suggested that the reason for his not publishing was that he had not completed his work or that he had feared that his ideas would be judged heretical by the religiously orthodox.

It is possible that such motivations played a role. It is also possible that Darwin was afraid that he might harm his reputation if he brought out a controversial theory before he had amassed the evidence that he needed. Even before *Origin of Species* was published, Darwin was considered one of the most eminent scientists of his day. He might have feared that his theory of natural selection would be interpreted as evidence of premature senility.

If such thoughts ran through Darwin's mind, they were all for nought. As it turned out, the theory of natural selection was accepted quite readily by the scientific community and, surprisingly, by the general public as well. To be sure, there were some scientific holdouts, and attacks were organized by members of the creationist camp. But the majority of the readers of Darwin's book found themselves convinced.

This quick acceptance may have had something to do with the fact that the Victorian idea of "progress" was very much in the air at the time. The Industrial Revolution had brought numerous technological advances, and many people believed that progress had been made toward a more just society as well. The mid-nineteenth century knew nothing of the modern problems of pollution, limited resources and preservation of the environment. At least they had not become the pressing concerns that they are today. If human society could progress, it seemed natural to believe that "progress" could be made in nature also.

But if the idea of evolution happened to fit in with prevailing notions, this was only one of the reasons for the favorable reaction to Darwin's book. The evidence that Darwin had accumulated was at least as important a factor. Unlike Wallace, and unlike early evolutionists such as Matthew, Darwin did much more than suggest that natural selection caused species to evolve. He had spent years studying the effects of artificial selection on domesticated plants and animals. He carefully outlined the evidence that he had gathered on the topic, suggesting that the effects of artificial and natural selection were similar. Then he went on to discuss such subjects as the evolution of instincts, fossils and the geological record, hybridism, morphology, embryology and the geographical distribution of living species. Not only did he show that the theory of natural selection was very plausible, he also presented so much evidence for evolution that it was no longer possible to doubt that it had taken place.

Paradoxically, although Darwin's theory proved to be correct in its broad outlines, it was wrong in many of its details. Since nothing was known of genetics in Darwin's day, he could not adequately explain the origin of the variations that would make natural selection possible. Since he knew nothing of genes or of mutations, he finally fell back onto a Lamarckian theory of inheritance of acquired characteristics. In 1865, six years after the publication of *Origin of Species*, he developed a rather outlandish theory called *pangenesis*. According to this hypothesis, the cells of the body produced minute corpuscles called *gemmules* which could affect cells in the reproductive organs. Darwin was not repudiating natural selection when he proposed this. He simply believed that a Lamarckian process produced the variation upon which natural selection acted.

Darwin also believed that variations had to arise simultaneously in the majority of the members of a species. Otherwise, he thought, the variations would be swamped out of existence when an animal bred. Today we know that a single organism can transmit a gene to innumerable offspring, and that, if the gene enhances the probability of survival, then it can easily spread throughout an entire population.

Finally, Darwin believed that evolution was always a very gradual process. Even this idea is being questioned today. Scientists have accumulated evidence indicating that, at least some of the time, evolution can proceed by fits and starts. It has become apparent that organisms can retain the same evolutionary form for many millions of years and then change very rapidly. Recent studies of fresh-water snails and bivalves, for example, have revealed that while some species look exactly the same as they did millions of years ago, others have undergone transformations in periods of five thousand to fifty thousand years.

Darwin should not be judged too harshly for his errors. Given the evidence that was available at the time, it is doubtful that anyone could have come up with a better theory. In fact, many of Darwin's speculations have turned out to be astonishingly accurate. He suggested, for example, that the human race evolved in Africa. We now know that he was correct. He anticipated the modern discipline of sociobiology by speculating about the evolution of altruistic behavior. And he made a statement that many modern primatologists would endorse. "He who understands baboon," Darwin jotted down in one of his notebooks, "would do more toward metaphysics than Locke."

When Darwin writes about the human races he sounds more like a modern anthropologist than like a nineteenth-century scientist. In his 1871 book *The Descent of Man*, he emphasizes the similarities, rather than the differences, between the races. He suggests that it is "our nice powers of discrimination gained by the long habit of observing ourselves" that make them seem dissimilar. He points out that he once observed that "a full-blooded negro with whom I happened to be intimate" possessed a mind that was really very similar to his own, and he states that he is "deeply impressed with the close similarity between men of all races in tastes, dispositions and habits." He does not fall victim to the misconception, common in his day, that the languages of primitive peoples were simple and animallike; on the contrary, he realizes that they can be very complex.

Darwin did speak of the "lower races," however, and he did seem to believe that, at least in some respects, white Europeans were the more advanced. Nevertheless, his comments on the topic of race seem quite enlightened when compared with those of his contemporaries. In Darwin's day, many scientists went so far as to assign whites and blacks to different species.

Today many anthropologists are of the opinion that the concept of race should be discarded altogether. Genetic studies have shown that the different racial groups are really quite similar to one another. It has been demonstrated that about 84 percent of all genetic variation consists of differences between individuals of the same tribe or nationality, while only about 10 percent of the total biological diversity arises from divergence between racial groups. In other words, if white and black individuals are chosen at random, the differences between them will be only slightly greater than the differences between two randomly chosen Frenchwomen, or between two members of the same tribe in Africa. It is true that differences in skin color are very conspicuous. However, skin pigmentation is controlled by a relatively small number of genes, perhaps as few as four. The total number of genes possessed by every human being is something of the order of a hundred thousand.

During the second half of the nineteenth century, few scientists were very interested in the similarities between the various racial groups. For the most part they ignored Darwin's relatively liberal view and allowed themselves to be influenced by the predilections of the age. Thus they rarely attempted to view the question of race in an objective manner. Instead, they tried to prove that the "lower races" were evolutionarily less advanced than the white one.. Darwin's ideas about race prompted little or no dispassionate inquiry. Instead, his evolutionary theories were used to prop up existing racist beliefs.

In general, the nineteenth-century scientists did not adhere to Darwin's (and Locke's) view that human beings were pretty much alike. On the contrary they held, with Galton, that there were important innate differences between individuals. When they wrote about racial differences, they expressed this view with a vengeance. The French anthropologist Pouchet spoke of "races placed so low that they have quite naturally appeared to resemble the ape tribe." Pouchet urged his readers to "descend boldly the steps of the human ladder"; he painted pictures of primitive peoples who lived in a state of "moral brutality"; and he attempted to prove that "certain races are not a whit more intelligent than certain animals." The Austrian physician and scholar Carl Vogt asserted, meanwhile, that "the pendulous abdomen of the lower races . . . shows an approximation to the ape." Blacks, he added, had feet that made "a decided approach to the form of a hand." As a result they could rarely stand quite upright. Vogt commented also on the evolutionary differences between men and women, concluding that "whenever we perceive an approach to the animal type, the female is nearer to it than the male."

The English scientists did not allow themselves to be outdone. They spoke of the "imperfect brow" and "projecting lower jaw" that were supposedly observed in blacks, and suggested that blacks and orientals were "human saurians." There arose a school of "degenerationists" who suggested that evolution sometimes ran backward, producing races that had sunk into savagery from some previously attained higher state, becoming "swamps and backwaters of noble humanity," rather than "representatives of the fountainhead from which it had been derived."

If scientists were to continue to place the "lower races" somewhere between the apes and white Europeans on the evolutionary scale, they needed a theory to justify their conclusions. But the idea of degeneration wasn't entirely adequate. It was more of an outgrowth of the Christian idea of the Fall of Man than a result of evolutionary theory. Although even Darwin believed in the existence of isolated *atavisms*, or "throwbacks," no one had really come up with any evidence that evolution could really run backward.

The sought-for theory was provided by Ernst Haeckel, the German zoologist and popularizer of Darwinian ideas. Haeckel resurrected a pre-Darwinian idea known as the *biogenetic law*. According to this "law," there were parallels between the embryonic development of an individual and stages in the fossil record. This was an idea that had first been proposed by certain romantic philosophers, who maintained that man was a microcosm in which all life was reflected. Animals were merely fetal stages of man.

Haeckel appropriated this idea and cast it into a more modern form. Pointing out that the human embryo had gill slits at one stage of its development, and a rudimentary tail at another, Haeckel proclaimed that "ontogeny recapitulates phylogeny." In other words, an individual, during its development (ontogeny), passed through stages that represented the evolution of its species (phylogeny). The human embryo resembled the adult form of an ancestral fish at one stage. At another, it was like a reptile. At yet another, it was similar to a primitive mammal.

Scientists repudiate the doctrine of recapitulation today. They point out that the human embryo passes through stages that resemble the embryos of ancestral species, not their adult forms. The reason for this is fairly simple. Since, for the most part, natural selection operates on individuals only after they are born, embryos tend to retain the appearance that they had millions of years ago. Naturally, this observation applies only to the early stages of development. A four-month fetus looks more like a human baby than like a fish or a reptile, or anything else.

However, the doctrine of recapitulation was quite influential during the latter half of the nineteenth century. It dominated the thought of the biologists and paleontologists who were attempting to reconstruct evolutionary lineages, And it furnished a "scientific" justification for the relative ranking of the various human races.

If earlier stages in an individual's development resembled the adult forms of primitive ancestors, the recapitulationists said, then it was obvious that the adult members of the "lower" races resembled children of the superior ones. In effect they were saying that the evolution of blacks had been arrested at a stage that corresponded to childhood in whites. Intellectually and emotionally, blacks were like children.

Some scientists extended the theory even further and looked for childlike characteristics in other "inferior" groups—for example, women and lower-class whites. More often than not, when they had decided what "facts" were needed, these scientists proceeded to find them.

Some scientists refused to accord women and blacks a status equal to that of white male children. The Swiss-born American zoologist Louis Agassiz compared the brains of adult blacks to those of seven-month white fetuses, while the French sociologist Gustave Le Bon claimed that the brains of many women were more like those of gorillas than "the most developed male brains." Women, Le Bon added, were "inferior forms of human evolution," and were "closer to children and savages than to an adult, civilized man."

Even Darwin ranked women below men on the evolutionary scale. He believed that women showed qualities that were characteristic of "a past and lower state of civilization." He apparently did not realize that he was identifying biological with cultural evolution. The confusion was quite common at the time.

Darwin attempted to explain the supposed inferiority of women by supposing that "half-human male progenitors of man" had had to make extensive use of their wits when competing for possession of females, and in defending their families against enemies. The women, who stayed at home and looked after the children, hadn't exercised their intelligence as much. As a result, their minds had evolved more slowly.

Incidentally, a variation of this theory is still encountered today. For many years it was customary to explain the rapid evolution of the human brain by supposing that male hunting produced evolutionary pressures that favored an increase in intelligence. Although the inference that women were therefore less intelligent was dropped, the argument was really not very different from the one that Darwin gave. In either case, one began with the assumption that the evolution of human intelligence could be traced back to the things that the *men* were doing long ago in the evolutionary past. Female activities were seen as being relatively unimportant.

Another idea that attracted a great deal of attention during the nineteenth century was the concept of atavism. It was believed that some individuals were evolutionary throwbacks. According to the Italian physician Cesare Lombroso, "born criminals" inevitably exhibited atavistic traits. There were marked with anatomical signs of their apishness and could be identified by certain physical stigmata. Lombroso compared criminals to savages. The lower races, he insisted, also had apelike features. Furthermore, criminality was normal behavior among the less-evolved peoples. Lombroso made much of the flattened nose of blacks, comparing it to the nose of monkeys. (He apparently never noticed that the thin lips of whites are more apelike than the thick ones of blacks.)

Lombroso believed that epilepsy was a mark of criminality, and claimed that every "born criminal" suffered from the disease to some degree. This particular theory had quite an influence within the eugenics movement; in the early twentieth century the eugenists commonly held that epileptics should be denied the right to breed.

Lombroso was not the only scientist who made use of the concept of atavism. Many scientists, including the great English biologist Thomas Henry Huxley, believed that the Neanderthal fossils, which had been discovered in 1857, were the remains of a "degenerate" modern individual. Although Darwin correctly pointed out that the Neanderthal skull was "well-developed and capacious," many years were to pass before it was recognized that Neanderthal man was not an atavistic creature, but a possible ancestor of modern man.

In England, meanwhile, the physician John Langdon Haydon Down put forward a theory that idiots were atavistic individuals. He found that many congenital idiots exhibited features that were supposedly characteristic of the lower races. Some white idiots were of the "Ethiopian variety." He described these as "white negroes, although of European descent." Others resembled Malaysians or American Indians. And there were yet other types. "A very large number of congenital idiots are typical Mongols," Down observed. Describing one such "Mongolian idiot" in detail, he stated that "there can be no doubt that these ethnic features are the result of degeneration."

Down thought of himself as a racial liberal. Perhaps he was, by the standards of the day. Like Darwin, he argued against the theory—then commonly held—that the various human races were actually separate species. Down felt that the fact that white idiots apparently resembled the adults of "lower" races was evidence for the unity of the human species.

Although modern scientists realize that Down's theories rested on nineteenth-century prejudice rather than on scientific fact, one of the items of terminology that Down invented still persists. Individuals who suffer from a certain type of birth defect are still frequently referred to as *mongoloids*, or *mongolian idiots*. The term *mongolism* is gradually being replaced by *Down's syndrome*, but the reform is not yet complete.

The scientists of the Victorian age were blatant racists. But perhaps it would be naïve to expect them to have been anything else. Racist ideas were almost universally held in those days. In this respect, the white Europeans were not very different from the members of the numerous other cultures that have existed on this planet. The ancient Greeks and the ancient Chinese also looked down upon the "inferior" barbarians. Similar views have been held by many civilized peoples. The members of primitive cultures are no different. Often they are more extreme. There still exist primitive peoples who consider the members of other tribes to be less than human. If the Victorians were racist, they were simply behaving as human beings have behaved in many times and in many places.

It is somewhat more difficult to forgive the Victorians for distorting scientific fact in order to justify their racial prejudices. Rather than study racial differences in a reasonably objective manner, they propounded outlandish theories that seemed to make the differences seem much greater than they really were. Many of the scientists of the day proceeded in an anything-but-dispassionate manner. They conjured up scientific "facts" that would support their theories and ignored evidence that might have suggested that they alter their views.

However, we must remember that only some of them did this. The great majority simply went on with their own pursuits. They didn't protest against the racism around them, but no one else did in those days either. If we are tempted to blame them for their acquiescence, it would be well to remember that many twentieth-century scientists have been equally guilty; scientific antiracism dates only from the end of World War II.

By now we should be prepared to give up the idea that science is a completely objective activity and that it operates independently of the predispositions that are characteristic of any given era. Scientists have just as many preconceived ideas as the rest of us. And whenever they turn their gaze upon their fellow human beings, those ideas are likely to play an important role.

3

The Survival of the Fittest

When John D. Rockefeller was battling for control of the oil-refining industry during the last decades of the nineteenth century, he engaged in tactics that were anything but ethical. Rockefeller's Standard Oil Company not only demanded that railroads give rebates if they wanted its voluminous business, it also forced them to pay a portion of the rates collected from other refiners. All this was done, not to increase profits, but to drive competitors out of business.

When these tactics were not effective, others were used. Standard Oil officials penned memos like the following: "Wilkerson & Co. received car of oil Monday 13th. . . . Please turn another screw." On at least one occasion, Standard Oil arranged to have a small explosion take place in a competing refinery. And, of course, it made numerous offers to buy competitors out. Often these were offers that could not be refused. "If we did not sell out . . . we would be crushed out," was the way that one competitor put it.

Rockefeller apparently had little difficulty reconciling his unscrupulous business conduct with his image of himself as a God-fearing man. He rarely missed a week teaching his Baptist Sunday-school class. In a much quoted Sunday-school address he stated:

> The growth of a large business is merely a survival of the fittest. . . . The American Beauty Rose can be pro-

> duced in the splendor and fragrance which bring cheer to its beholder only by sacrificing the early buds which grow up around it. This is not an evil tendency in business. It is merely the working-out of a law of nature and a law of God.

When Rockefeller spoke of "survival of the fittest," he was not echoing Darwin; he was making use of a phrase that had been coined by Herbert Spencer some nine years before *Origin of Species* was published. He was not referring to Darwin's idea that species evolved by natural selection, but rather to Spencer's doctrine that evolutionary ideas could be applied to human societies.

Spencer is not much read today. Histories of modern philosophy sometimes fail even to mention his name. Yet in his own day, he was perhaps the most widely read of all philosophers. He was read, not only by businessmen, but also by doctors and lawyers, by scientists and by clergymen, even by laborers. He was admired by eminent naturalists and by members of the general public. Alfred Russel Wallace called Spencer "the greatest all-round thinker and most illuminating reasoner of the Nineteenth Century." Darwin hailed Spencer as "about a dozen times my superior." On another occasion he called Spencer "by far the greatest living philosopher in England; perhaps equal to any that have lived."

It was in the United States that Spencer's ideas were most influential. His American publisher sold hundreds of thousands of copies of his books. Sociologists, economists, and novelists such as Jack London and Theodore Dreiser were influenced by his thought. The capitalist entrepreneur Andrew Carnegie counted himself as one of Spencer's disciples. The United States Constitution was, at times, interpreted by the Supreme Court according to Spencer's principles. The three Spencerians who sat on the Court were so influential that Justice Oliver Wendell Holmes was once led to remark, in a dissenting opinion, that "the Fourteenth Amendment does not enact Mr. Herbert Spencer's *Social Statics*."

Spencer did as much as anyone to popularize Darwin's theories. But, paradoxically, he was not a Darwinian. On the contrary, he believed in the Lamarckian inheritance of acquired characteristics. Although he accepted natural selection, he believed that it played only a secondary role. Evolution, according to Spencer, proceeded in a Lamarckian manner. At best, natural selection only speeded the process up. Spencer's views were thus almost diametrically opposed to those of Darwin, who believed that natural selection was most important and used Lamarckian ideas only to explain the origin of the variations upon which natural selection operated.

Unlike Darwin, Spencer was not especially interested in biological evolution. He was not a biologist, and he did not emphasize the importance of evolution for explaining the origin of species. To Spencer, evolution was a grand philosophical idea that could be applied to practically everything. Spencer had little interest in attempting to solve any of the outstanding scientific problems. He wanted, instead, to create an all-embracing system that would provide a synthesis of the accumulated knowledge of the age. Spencer believed that, with the help of the principle of evolution, he could acomplish this task. In Spencer's writings, evolution ceases to be a scientific concept. It becomes a grand cosmic principle.

Spencer maintained that evolution was a law of nature that could be applied universally. Everything evolved from the simple to the complex. This idea could be applied to the surface of the earth. Since the earth had been formed, Spencer said, its surface had been growing more irregular and its climate more variegated. Human societies and language alike became more complex. So did the tools that human beings used. The principle could even be applied to sexual relationships. Originally, Spencer claimed, an unfettered promiscuity had been the rule. Over the course of time, it had gradually evolved toward the pinnacle represented by Victorian monogamy.

And, of course, the principle of evolution could be applied to the universe itself. Stars and planets, Spencer said, evolved from gaseous nebulae. Like everything that it contained, the universe too attained greater complexity over the course of time.

Spencer had no argument with the prevailing racism of his day. He accepted the common but mistaken notion that primitive peoples had simple languages and simple social systems. When he encountered evidence to the contrary, he attempted to explain it away. For example, he was aware of the complex kinship systems of the Australian aborigines. But he interpreted this as evidence that they had regressed from a higher state. Savages, Spencer said, were physically and intellectually inferior to European whites. He compared the members of certain native tribes to orangutans, and he embraced the recapitulationist doctrine that members of the "lower" races were like white children.

Spencer was less interested in biological evolution than he was in the evolution of the human mind. He believed that evolution shaped intellectual and moral characteristics alike. This allowed him to come to the conclusion that societies had to evolve also, as the individuals of which they were composed gradually evolved to higher intellectual and moral states.

According to Spencer's neo-Lamarckian theory, human evolution was brought about by human effort. Those individuals who strove to use their minds would pass the resulting increased intellectual capacity along to their offspring. Similarly, acquired moral character was a heritable trait. As individuals developed such qualities as "sociality" and altruism, cultures progressed from savagery, to barbarism, to modern societies that were governed by law.

Spencer attempted to develop a system of ethics that was based on these ideas. If one wanted to promote human happiness, he said, then nothing must be done to interfere with the evolutionary process. If evolution was allowed to follow its natural course, then progress would be the inevitable result.

In particular, Spencer went on, governments must not enact social legislation designed to alleviate the lot of the poor. By doing so, they interfered with the weeding out of the unfit, and increased the burden of misery that future generations would have to bear. The physically and intellectually feeble had to be eliminated if the amount of human happiness was to be increased. It was best that those "not sufficiently complete to live" should be allowed to die. It made no difference whether the "incompleteness" consisted of lack of "strength, or agility, or perception, or foresight, or self-control"; the fundamental law of nature had to be observed.

Spencer admitted that a certain amount of suffering would be caused by such policies. "It seems hard," he said in his 1850 book, *Social Statics,* "that a laborer incapacitated by sickness from competing with his fellows should have to bear the resulting privations. It seems hard that widows and orphans should be left to struggle for life and death." Nevertheless, he went on, we must learn to steel ourselves to such sights if we want to further the progress of humanity. The "purifying process" had to be allowed to run its course. Otherwise, the evils would multiply and posterity would have to labor under a continually increasing curse.

If "nature's failures" were not helped to propagate their kind, then the survival of the fittest would lead to the betterment of man and society alike. Spencer thought that just such a process was taking place in the United States. The varieties of the Aryan race which formed the population of that country, he said, would eventually produce "a finer type of man than has hitherto existed." As a result, America would eventually produce a civilization "grander than any the world has known."

The proponents of laissez-faire capitalism in the United States returned the compliment by popularizing Spencer's theories and applying them to the competitive struggle. The best competitors—the fittest—would win out. This, they claimed, would lead to a continuing improvement of society. If workers were forced to labor long hours for low wages and were subjected to dangerous and unhealthy working conditions, there was nothing wrong with that. After all, Spencer had shown that, if the battle for life was made more fierce, then evolution would progress all the faster. If men like Rockefeller drove their competitors out of business, they were only accelerating the betterment of society.

Social Darwinism is the name that has been given to this kind of thinking. It is not a very accurate term. Social Darwinism, after all, has little to do with Darwin's theories. But perhaps it was only natural that Spencer's ideas should have been thought of as "Darwinistic." The Americans had taken to Darwinism even more enthusiastically than the populace of England. For the most part, American scientists had been prompt in accepting the theory of natural selection, and the American reading public had become fascinated with evolutionary thought shortly after the end of the Civil War. If they failed to distinguish between the theories of Darwin and those of Spencer, they cannot be blamed for it. After all, at the time, Spencer was rapidly gaining a worldwide reputation as the philosopher of evolutionary thought.

Spencer's theories were taken up by scientists, economists, political leaders and businessmen alike. The principles of social Darwinism were proclaimed from pulpits, expounded in books and in lectures, and discussed in newspaper editorials. Capitalist entrepreneurs and business executives took to the doctrine enthusiastically. Naturally, they had little to criticize in a theory which "scientifically proved" not only that they were the most fit, but also that their underhanded practices were justified. When Spencer visited the United States in 1882, prominent men from American cultural, political and business life staged a banquet in his honor at Delmonico's in New York. Their tributes were so enthusiastic that even the rather egotistical Spencer found himself embarrassed.

Among those giving testimonials was the steelmaker Andrew Carnegie. It is hard to imagine that anything could have prevented Carnegie from participating in the tribute. He was one of Spencer's most enthusiastic American admirers. Years previously, when he had read Spencer's work for the first time, he had experienced something resembling a revelation.

Carnegie had lost his religious faith as a young man. For many years, he had failed to find a substitute. It was not long before he was making a lot of money—using methods that were not unlike those of Rockefeller—but he did not find this entirely satisfying. In 1868 he had written, for his own guidance:

> Man must have an idol. This amassing of wealth is one of the worst species of idolatry—no idol more de-

> basing than the worship of money. . . . To continue much longer overwhelmed by business cares and with most of my thoughts wholly upon the way to make more money in the shortest time must degrade me beyond hope of permanent recovery.

A few years later, Carnegie read Spencer. At once, his perplexity disappeared. This is how he was later to describe the experience in his *Autobiography:*

> I remember that light came in a flood and all was clear. Not only had I got rid of theology and the supernatural, but I had also found the truth of evolution. "All is well since all grows better," became my motto, my true source of comfort. Man was not created with an instinct for his own degradation, but from the lower he had risen to the higher forms. Nor is there any conceivable end to his march to perfection. His face is turned to the light; he stands in the sun and looks upward.

While Spencer's theories were providing Carnegie with a new faith, their wide acceptance was impeding social progress in the United States. Social Darwinism had an influence upon legislation, executive actions and court decisions alike. Naturally, Spencer's ideas were widely accepted within the Republican party. But the Republicans were not the only ones to come under Spencer's sway. For example, in 1887, Democratic President Grover Cleveland vetoed a bill that would have helped Texas farmers buy seed grain during a drought. Federal aid under such circumstances, he said, "weakens the sturdiness of our national character." It was a sentiment of which Spencer would have wholeheartedly approved.

During the closing years of the nineteenth century, a number of states began to enact various kinds of reform legislation. But much of it was struck down by the Supreme Court. Three of the Justices, Stephen Field, his nephew David Brewer, and Rufus Peckham, were avowed Spencerians. The decisions in which Field, Brewer and Peckham participated struck down legislation on hours of work, minimum wage and child labor as unconstitutional.

The social Darwinist philosophy was used to bolster conservative outlooks in a number of different ways. Only the evolutionary process, the conservative advocates of social Darwinism said, could improve society. Therefore one should do nothing to interfere with the competitive struggle. The idea that evolutionary changes must necessarily be slow and gradual also lent support to the conservative outlook by implying that one should do nothing to interfere with the status quo. Although it might take centuries, it was claimed, social evils would eventually disappear of themselves. Any attempt to legislate them out of existence would inevitably backfire, making the process that much slower.

Spencer himself was less of a social Darwinist than his disciples. In particular, he did not have a high opinion of American capitalism. When his admirers feted him at Delmonico's, he did not praise them for engaging so ferociously in the competitive struggle. On the contrary, he advised them to struggle less and to cultivate the art of relaxation. On another occasion, he wrote that "trade is essentially corrupt," and denounced the unethical business practices of the day. "A system of keen competition carried on, as it is," he said, "without adequate moral restraint is very much a system of commercial cannibalism." It is not surprising that, when Carnegie invited Spencer to Pittsburgh so that he could view capitalism at its most efficient, the English philosopher found the city and its steel mills to be "repulsive."

Still, Spencer was no liberal. He was as much an opponent of reform as his disciples in the United States. In fact, many of Spencer's ideas are still part of conservative political doctrine. Spencer was opposed to all social legislation and to all forms of government regulation. He believed that governments had only one justifiable purpose: the administration of justice. Everything else, including the delivery of mail and municipal sanitation, should be in the province of private enterprise. In some respects, he was even more conservative than present-day extremists. For example, he was opposed to public education and argued that government-run schools could too easily be used to indoctrinate.

Today such ideas are not thought to have much to do with evolution. But they were in Spencer's day. Spencer's speculations had led him to a theory of human nature. He believed that humanity was engaged in a struggle for existence, and that evolution weeded out those who were inferior. Like Galton, he was convinced that there were significant hereditary differences between human beings, and that the "unfit" should not be encouraged to propagate their kind. Human nature, in Spencer's view, was not something that was molded by environment. On the contrary, human character was inherited, and was molded by the battle for existence.

Spencer believed that great benefits would accrue to humanity if nothing was done to interfere with this struggle. This implied that the ideal economic system was laissez-faire capitalism, and that conservative political and economic policies were intrinsically the most just.

But others saw nothing in evolutionary theory that necessitated such an outlook. Before the social Darwinists had even begun to preach their doctrines, Karl Marx was using evolution as a justification for socialism. After reading Darwin's *Origin of Species* in 1860, Marx declared that "Darwin's book is very important and serves me as a basis in natural science for the class struggle in history."

Marx was later to elaborate upon this idea in a letter to his friend and colleague Friedrich Engels. "It is remarkable," Marx said, "how Darwin recognizes among beasts and plants his English society with its division of labor, competition, opening up of new markets, 'invention,' and the Malthusian 'struggle for existence.'" Engels did not fail to pick up the cue. In 1876 he wrote an essay called "The Part Played by Labor in the Transition from Ape to Man."

In Germany, socialist bookstores displayed the works of Marx and Darwin side by side. Meanwhile, American socialist intellectuals also made use of evolutionary concepts, and even quoted from

Spencer in their writings. They professed to be indebted to Spencer for pointing out that societies could evolve in a manner similar to biological species. When the American socialist Arthur M. Lewis published a book titled *Evolution, Social and Organic*, no one thought it odd that he should see a relationship between science and revolution; the book had the largest advance sale of any American socialist publication.

The liberal American sociologist Thorstein Veblen also made use of evolutionary ideas when making observations about human nature. In his influential book *The Theory of the Leisure Class*, published in 1899, Veblen launched a wry, often ironic attack upon the capitalist entrepreneurs. Unlike the social Darwinists, he did not consider them to be the "fittest." On the contrary, he drew upon the theories of Spencer and Lombroso, and characterized them as throwbacks to an earlier, barbarian, stage of evolution.

The entrepreneurs, Veblen wrote, exhibited "predatory aptitudes and propensities carried over by heredity and tradition from the barbarian past of the race." In this respect they resembled lower-class criminals. The competitive struggle, in Veblen's eyes, did not lead to human improvement. It did just the opposite; it tended "to conserve the barbarian temperament, but with the substitution of fraud and prudence, or administrative ability, in place of that predilection for physical damage that characterizes the early barbarian." The capitalist businessmen, in other words, were not the evolutionarily most advanced; they were atavistic throwbacks who retarded social progress.

The iconoclastic Veblen may have delighted in his ability to stand social-Darwinist theories on their head. But other liberal and radical authors tended to steer away from the doctrine that there were significant innate differences between human beings. They might use evolutionary ideas to justify their faith in the possibility of social progress. But when they spoke of human nature, they tended to throw aside the kind of outlook that was characteristic of Galton and Spencer and their disciples, and to replace it with a thoroughgoing environmentalism.

One such author was John Peter Altgeld, who was later to be elected governor of Illinois and then to be voted out of office after one term when his constituents became aware of his radical sympathies. In 1884, Altgeld published *Our Penal Machinery and Its Victims,* in which he argued that poverty, not hereditary criminal nature, was the cause of most crime.

This book had quite an influence upon the lawyer Clarence Darrow, among others. After reading it, Darrow made a trip to Chicago to shake the hand of the author. Darrow was to spend the next forty years of his life propagandizing its main point, that crime resulted from the unjust conditions of human life. Darrow expressed this in an especially cogent way in a speech that he once made to a group of inmates of the Cook County penitentiary. "If every man, woman and child in the world had a chance to make a decent, fair, honest living," he said, "there would be no jails and no lawyers and no courts."

Yet Darrow was no antievolutionist. In fact, he exhibited quite a good understanding of Darwinian theory when he defended biology teacher John T. Scopes at the 1925 "monkey trial" in Dayton, Tennessee. According to the accounts that have been given of the trial, Darrow practically crucified the creationist political leader William Jennings Bryan when the latter took the stand.

Like the teachings of St. Paul, the concept of human evolution has been "all things to all men." It has been used to justify socialism and laissez-faire capitalism, racism and the equality of the human races, imperialism, nationalism and pacifism. To a much greater extent than any other scientific theory, it has been interpreted in a multitude of different ways. It seems that, whatever the preconceptions that one entertains, it is possible to find an evolutionary argument to support them.

The problem is this: both hereditarian and environmentalist outlooks seem to be reasonably consistent with Darwinist theory. Everyone agrees that some human behavior has a biological foundation, and that some does not. But it is not easy to tell what the most important hereditary qualities are. Is intelligence one of them? Galton believed that intelligence was the result of heredity.

Darwin believed that, intellectually, human beings were pretty much alike and that significant scientific discoveries were to be attributed more to hard work than to any innate "genius." Which man was right? Even today that question is not so easy to answer. The debate over the significance of I.Q. tests has not yet been settled. Although it has been shown that I.Q. has a heritable component, not everyone agrees that the tests really measure "intelligence." Some psychologists maintain that intellectual ability is a multifaceted thing that cannot possibly be measured by a numerical score. Others take the point of view that intelligence can be *defined* to be I.Q.

Obviously, musical talent can be inherited. If it could not be there would be no prodigies like Mozart. But what about scientific or mathematical ability? Or literary talent? No one really knows.

One thing that is clear is that environmentalism tends to be associated with liberal political outlooks, while the hereditarian viewpoint is taken up by conservatives. Liberals tend to associate social ills with the poor environments to which some individuals are subjected. Like the disciples of Locke, they tend to think that human nature will be improved if we can only institute a better society. The way to reduce the amount of criminal behavior, they say, is not to lock up criminals for longer periods, but to modify environments that make people criminals in the first place.

Conservatives, on the other hand, more often think of human nature as something that is inherited and unchangeable. They are more likely to believe that one can be a criminal "by nature." They are less likely to believe that social change can be legislated; they resist the notion that human beings can be forced into new molds.

Generalizations have a way of doing violence to political viewpoints, so perhaps it would be better if I didn't continue in this vein. But I feel that the generalizations that I have made so far are reasonably accurate. The environmentalism of the liberal Darrow and the hereditarian viewpoints of the conservative Spencer are not unlike outlooks that are encountered today.

It is not my intention to argue that one kind of political outlook is superior to another by reason of its being more consistent with a scientific view of human nature. I am concerned only with demonstrating that politics and science can become intertwined. As we shall see in a subsequent chapter, they are still intertwined today. For example, political arguments have been used by the critics of sociobiology.

But perhaps it would be best if, at this point, we didn't attempt to look too far ahead. It might be better to continue with the story of how science, in its blundering way, finally managed to unravel some of the mysteries of human evolution.

♣ 4

Missing Links

In *The Descent of Man,* Darwin makes a very revealing comment concerning the fossil human skulls that were known at the time. "Considering how few ancient skulls have been examined in comparison with recent skulls," he says, "it is an interesting fact that in at least three cases the canines project largely; and in the Naulette jaw they are spoken of as enormous."

This is a curious observation. The Neanderthal jaw that was discovered in 1866, in a limestone cave in Trou de la Naulette, Belgium, does not have large canine teeth. As a matter of fact, it is toothless. It lacked teeth when Darwin wrote about it, and it lacked teeth when it was found.

But Darwin cannot be criticized for making his remark. He was only quoting prominent scientists who had subjected the jaw to careful scrutiny. Although the teeth were missing, these men had no doubts that the "enormous" canines had once been present. After all, gorillas had large canines, and it was obvious that the fossil was, as Darwin put it, "directly interposed between the jaw of an anthropoid ape and a modern human being." The jaw, in other words, was thought to be the fossil of a missing link.

The scientists of the 1860s understood very well that man was not descended from any living ape. Nevertheless they did not hesitate to conclude that this jaw, which seemed quite massive and was marked by a number of morphological peculiarities, such as the absence of a projecting chin, came from a creature that was half man and half ape. It was just the kind of intermediate form that they had expected to discover.

The idea of a "missing link" is really a pre-Darwinian concept. The ancient Greeks had recognized the affinity between human beings and apes, and they had speculated about the possibility of intermediate forms. This speculation continued intermittently through the ages. During the eighteenth century the pace of speculation quickened. Philosophers began to discuss the idea of a Chain of Being. According to this theory, all beings, from plants to animals, to human beings, to angels, could be assigned places on an ascending scale of perfection. Mammals, for example, were placed above fish and reptiles, and men were placed above all other earthly creatures, including women.

There was one very obvious gap in the Chain between human beings and apes. Many philosophers had no doubt that this "missing link" existed somewhere. Thus when early voyagers began to encounter primitive peoples, they did not hesitate to assign the latter a place in the gap. Among the first missing links were the Hottentots of Africa. It was said that Hottentots had an apelike appearance, and that their language was only slightly more sophisticated than that of chimpanzees.

When the Australian aborigines were encountered, they too were classified as beings who were less than human. During the latter part of the nineteenth century, scientists made much of the supposed similarities between the Australians and the Neanderthals. Until around the middle of the twentieth century, textbooks of anthropology frequently stated that the aborigines constituted the most primitive of the races of man.

But "primitive" races were clearly not missing links in the evolutionary sense. Although nineteenth-century scientists believed that blacks occupied a lower place on the evolutionary scale than European whites, they did not believe that the latter were descended from the former. However, they had every expectation that the evolutionary links between man and the apes would quickly be found. Although only a few human fossils had been discovered by the 1860s, the paleontologists of the day did not doubt that the evolutionary record would soon become complete. Convinced that man had evolved over a period of tens of thousands of years (today we know it took millions), they thought that they would quickly discover a series of intermediate forms that would take them all the way back to man's apelike ancestors. When the fossils that they did find turned out to be not very apelike at all, these scientists became victims of their own imaginations. They saw simian features where none existed. In the case of the Naulette jaw, they made a missing link out of a fossil of *Homo sapiens*.

Today, Neanderthal man is classified as *Homo sapiens neanderthalensis*, a subspecies of *Homo sapiens*. The full scientific name for modern man is *Homo sapiens sapiens*. In each case, the second word designates the species, and the third the subvariety.

Neanderthal man first appeared in Europe about 125,000 years ago. He was adapted to cold, glacial climates. Like the modern Eskimo, he was short and stocky. The average specimen stood about five feet tall and weighed approximately 160 pounds. Neanderthal had a brain that was slightly larger than the modern average. Studies of Neanderthal skulls have revealed that the average cranial capacity was about 1,500 cubic centimeters. The modern average is 1,370.

One should not conclude that Neanderthal was more intelligent than modern man, however. Brain size is not well correlated with intelligence. The normal range for modern populations is 950 to 2,200 cubic centimeters, and eminent individuals have had brains that approached both the low and the high limits. For example, the Nobel prize-winning French author Anatole France had a brain that measured only 1,017 cubic centimeters, while that of the great Russian novelist Ivan Turgenev measured 2,012.

There are some striking differences between Neanderthal and modern skulls. Neanderthal had a prominent bony eyebrow ridge, for example, and his skull was longer and flatter than that of a modern human. But no one knows what, if any, significance should be attached to the difference in brain shape. It is possible that the

Neanderthals may have been slightly inferior to moderns in intelligence. On the other hand, they might well have been our intellectual equals, or possibly even our superiors.

Neanderthal did not possess the gorillalike fangs that Darwin attributed to him. In fact, he looked very much like modern man. If one could take a Neanderthal man, give him a shave, dress him in a suit, and send him walking down the street, it is unlikely that he would attract any special comment or attention. The only really striking thing about his appearance would be his short stature.

Neanderthal had a culture that resembled that of modern primitive societies. He made and used tools, he buried his dead, and he took care of the old and disabled. One Neanderthal skeleton is the remains of an individual who had an arm that had been amputated at the elbow, who was crippled in one leg, and possibly blind in one eye. This individual died when he was buried by rocks from the collapsing roof of a cave. Studies of the skeleton indicate that his injuries were sustained long before his death; apparently his people cared for him for years.

Neanderthal is not the apelike "caveman" that late-nineteenth- and early-twentieth-century scientists thought him to be. He did not walk with a slouching gait; on the contrary, he stood fully erect. He may not have been especially hairy (hair is not preserved in fossils, so no one can be sure), and there is no convincing evidence that he was driven into extinction when modern man appeared in Europe about 35,000 years ago. Neanderthal may simply have evolved into the modern form. Alternatively, he may have interbred with *Homo sapiens sapiens*, losing his distinctive Neanderthal characteristics in the process.

Neanderthal was made into a "missing link" because a missing link was what nineteenth-century paleontologists required. These scientists described him, not as he was, but as they wanted him to be.

The story begins in 1856. In that year, two workmen came across a fossilized human skeleton in a quarry in the Neander

Valley* in western Germany. They promptly discarded the bones, tossing them out of the cave in which they had been digging. However, a foreman gathered up the skullcap and a few of the postcranial bones (*postcranial* is a term used to describe everything from the neck on down) and took them to a schoolteacher in nearby Elberfeld. The teacher, Johan Karl Fuhlrott, recognized that the bones had features that distinguished them from modern specimens, and he took them to Herman Schaaffhausen, an eminent anatomist who lectured at the University of Bonn.

The following year Fuhlrott and Schaaffhausen described the specimen in a paper that they presented to the Lower Rhine Medical and Natural History Society. They asserted that it was the fossil of a primitive man who had lived in Germany long ago. They concluded that the brow ridges and thick bones were racial characteristics. The remains, the author said, were those of "an individual of a savage and barbarous race derived from one of the wild races of northwestern Europe, spoken of by Latin writers."

Fuhlrott and Schaaffhausen presented their findings in 1857, two years before the publication of *Origin of Species*. At the time, scientists had not yet begun to speculate about fossil ancestors of modern man, and many of them refused to believe that the bones were as old as the two Germans suggested. One anatomist dismissed the find as the remains of an "old Dutchman," while one of his French colleagues maintained that they resembled the bones of a modern Celt, such as a "modern Irishman with low mental organization." A Professor F. Mayer of Bonn constructed a more elaborate theory. The bones were those of a Mongolian Cossack who had been riding through Prussia in 1814 in pursuit of Napoleon's fleeing army. The peculiar shape of some of the bones, he said, could be attributed to a case of advanced rickets. Obviously the poor man had died of his wounds while taking refuge in the Neander Valley,

*In German, "Neander Valley" was *Neanderthal*. In modern German, the *h* in *Thal* ("valley") is dropped. Thus, *Neandertal* has become an alternate spelling. However, the *h* is always retained in the scientific name *neanderthalensis*.

and a flood had washed his bones into the cave.

If Fuhlrott and Schaaffhaussen had been able to determine how old the bones were, both they and their critics would have been surprised. But at the time, the only methods available for establishing the age of fossils were very crude. One of these was the tongue test. It was thought that fossils of great antiquity would adhere to the tongue, while younger specimens would not. Another test consisted of dissolving a fragment of bone in hydrochloric acid and then looking to see if any gelatine remained.

Although both tests were eventually discarded as practically useless, they did have some scientific foundation. When a bone is fossilized, the organic material that it contains is gradually replaced by minerals. A fossil is really nothing more than a rock which is molded in the form of a bone that no longer exists. An old fossil which has lost all traces of an organic material called *collagen* does have a tendency to stick to the tongue. However, this is a somewhat subjective method of determining age. The hydrochloric-acid test wasn't very reliable either, at least not until it was resurrected in another form in modern times.

Since Schaaffhausen could not prove that the bones were really as old as he claimed, the scientists of the day felt justified in dismissing his claims concerning their antiquity. If, just a few years after the Neander Valley discovery, scientists had not begun to engage in debates about human evolution, the specimens most likely would have been forgotten. It is probable that their fate would have been similar to that of a Neanderthal skull that was found on Gibraltar in 1848.

The Gibraltar skull seems to have turned up during the repair of some fortifications. Some sixteen years passed before the governor of the military prison presented it to the anatomist George Busk, who displayed it at a meeting of the British Association for the Advancement of Science. Apparently it didn't evoke much interest, for it was stored away in the museum of the Royal College of Surgeons, where it remained until the beginning of the twentieth century.

When a fossil jaw was discovered at Trou de la Naulette in 1866, however, times had changed. Scientists had begun to search for a missing link. As a result, the jaw evoked an entirely different kind of reaction. Although years were to pass before it was classified as a Neanderthal specimen—in 1866, the term *Neanderthal man* had not yet been coined—the scientists of the day did realize that the jaw was not of recent origin.

Some of the anatomists who studied the jaw imagined that it was intermediate in form between the jaw of an ape and that of a modern human being. So certain were they of their interpretation that they did not hesitate to state that it must once have contained "enormous" canines (these are the canines Darwin spoke of). According to the French anthropologist Paul Broca, the Naulette jaw was "the first evidence to provide the Darwinians with an anatomical argument." It was, Broca went on, "the first link on the chain which, according to them, ought to lead from man to monkeys."

And then, in 1886, two skeletons that had cranial features similar to those of the Neander Valley specimen were discovered in the Spy caverns of Belgium. The bones were found amid fossils of various extinct Ice Age mammals, such as the woolly rhinoceros and the mammoth. The antiquity of Neanderthal man was finally established.

So, the Neanderthal fossil that had been dismissed as an "old Dutchman" or a Mongolian Cossack now, in 1886, attained the status of a missing link. As we have seen, the two views were equally incorrect. Although Neanderthal could very well be an ancestor of modern man, he is certainly not intermediate between *Homo sapiens* and a monkey.

Neanderthal man provides a classic example of the way in which theory can determine how "facts" are interpreted. Almost as soon as scientists began to speculate about evolutionary missing links, a missing link was found. It made no difference that Neanderthal was a poor candidate; scientists simply distorted the evidence to fit their preconceptions.

The scientists of the day were not disturbed when it became

apparent that Neanderthal had quite a large brain. It was not the size of the brain that was important, they decided, but rather Neanderthal's receding forehead. In the late-nineteenth century, the prefrontal lobes were often taken to be the seat of intelligence. Since the prefrontal area in the low-browed Neanderthal skulls was small, it was concluded that he had a low, "brutish" intellect.

Paradoxically, this conclusion was reached at the very same time when eminent anatomists were measuring the total brain capacities of blacks and of women in an effort to prove that they were intellectually inferior to white males. They apparently saw no contradiction in the assumption that brain size was a measure of intelligence in women and in blacks, while it was the prefrontal area that was important in the case of Neanderthal fossils.

The downgrading of Neanderthal man continued well into the twentieth century. In 1924 the English anatomist Grafton Elliot Smith characterized Neanderthal as "uncouth and repellent," describing him in the following terms:

> His short, thick-set and coarsely built body was carried in a half-stooping slouch upon short, powerful and half-flexed legs of peculiarly ungraceful form. His thick neck sloped forward from the broad shoulders to support the massive flattened head, which protruded forward, so as to form an unbroken curve of neck and back, in place of the alternation of curves which is one of the graces of the truly erect *Homo sapiens*. The heavy overhanging eyebrow-ridges and retreating forehead, the great coarse face with its large eye-sockets, broad nose and receding chin, combined to complete the picture of unattractiveness, which it is more probable than not was still further emphasized by a shaggy covering of hair over most of the body.

When Elliot Smith wrote these words, he was describing a reconstruction made by the French anthropologist Pierre Marcellin

Boule of a Neanderthal skeleton that had been found in a cave in the commune of La Chapelle-aux-Saints in 1905.

When Boule made the reconstruction and wrote a description of it, he was influenced by the scientific prejudices of the day. At the time, it was thought that Neanderthal had pronounced "simian" characteristics. Boule reconstructed the fossil in such a way as to support the prevailing theory. According to Boule, Neanderthal did not have a fully erect posture. His head was bent forward, and he walked with a bent-knee gait. His legs were so bowed that he walked on the outer edges of his feet, and he had a grasping big toe like an ape.

Boule stressed that Neanderthal had a "bestial" appearance; he had occupied "the lowest rung of the human ladder" and had possessed a brain that was intermediate between the brain of a chimpanzee and that of a "modern Frenchman." In his book *Fossil Men,* which was published in 1921, Boule repeatedly pointed out what he thought were similarities between Neanderthal and the chimpanzee, orangutan and gorilla. In Boule's opinion, Neanderthal was exactly what the late-nineteenth-century scientists had believed him to be.

This derogatory picture of Neanderthal was accepted until the middle 1950s. Then, in 1955, William L. Straus of Johns Hopkins Medical College and A.J.E. Cave of St. Bartholomew's Hospital Medical College in London subjected the La Chapelle skeleton to a reexamination. They found that the specimen upon which Boule had based his reconstruction had suffered from arthritis of the spine (a fact that Boule had never mentioned). As a result, he had had a bent posture that could hardly have been typical of a normal Neanderthal. Furthermore, there was nothing to indicate that the posture of Neanderthal was in any way different from that of modern man. There was no anatomical evidence to indicate that the Neanderthal knee could not be fully extended, and Neanderthal did not walk on the side of his feet.

Straus and Cave found that Boule had reconstructed the skeleton in a rather fanciful way. The specimen had originally existed in a fragmentary form, and quite a bit of guesswork had been necessary. But Boule had not carried out the job of reconstruction in a reasonable manner. For example, he had placed the center of gravity so far forward that, if Neanderthal had really looked like that, he would have fallen on his face every time he tried to take a step.

Today it is easy to see that Neanderthal's supposed subhuman attributes existed only in the minds of the scientists who studied him. This raises the question of why the picture of Neanderthal as a primitive, "bestial" creature persisted for so long a time. The nineteenth-century desire to find a missing link certainly played a role. For about thirty years, from the 1860s to the 1890s, the Neanderthal fossils were the only known specimens that seemed to throw any light on the problem of human evolution.

But this does not explain why this picture of Neanderthal should have persisted until the middle of the twentieth century. To be sure, the "old man" of La Chapelle (he was about forty when he died, which was old for a Neanderthal) was a pathological specimen. But he was not the only Neanderthal fossil known. Why was Boule's characterization of him not overturned sooner?

Could it be that scientists and lay people alike found something attractive in the idea that they had such a "bestial" forebear? Could they have been ascribing to Neanderthal all of the "animal" characteristics that they wanted to deny in themselves?

One hesitates to attempt to psychoanalyze an entire generation or to speculate what was going on in the subconscious minds of anthropologists who are now long dead. However, one cannot deny the fact that when Neanderthal was downgraded, modern human beings were elevated by comparison. Calling Neanderthal a "bestial" creature was a way of saying that civilized human beings had evolved beyond that. When one pictured Neanderthal as a creature with a dim mind and slouching gait, modern man seemed all the more "noble" by comparison.

The fact that the "cave man" image of Neanderthal entered the popular consciousness during the years immediately following World War I may not be as irrelevant as it seems. We too easily forget what a shock the First World War was to the Europeans of the era. The carnage that the war caused was unlike anything that had been known. In France, one person of every twenty-eight died. I am not talking about the casualty rate in the French army. I mean one person out of twenty-eight in the total population. In Germany, the figure was one out of thirty-two, in England one out of fifty-seven. In these countries, an entire generation of young men was decimated.

Among Americans, the death rates were relatively low. However, at the time, the United States was still a nation of immigrants. Many people had retained ties with relatives in Europe. Although Americans did not experience the full horror of the war, they knew that horrors were taking place.

Of course, this is all very speculative. But I can't avoid thinking that the picture of the short, hairy, burly man who carried around a club and dragged women into caves by their hair might have been consoling. The war had shown what savages human beings could be. Comparison with Neanderthal suggested that perhaps human nature wasn't so bad a thing after all; it had once been worse. It was a way of making the beast within us seem less threatening.

Boule completed his Neanderthal reconstruction in 1913. He hardly had a chance to write or lecture about it before the scientific world found itself embroiled in controversy over yet another "missing link." In 1912 the English solicitor and amateur geologist Charles Dawson and the distinguished English paleontologist Arthur Smith Woodward announced the discovery of Piltdown man. This was followed by the discovery of fragments of a second specimen in 1915. While Neanderthal continued to be the center of attention on the Continent, British scientists endlessly discussed the significance of their own fossil, the "earliest Englishman," as he was sometimes called.

Decades later, it would be discovered that Piltdown man was a fake, that he had been constructed from fragments of a modern human skull and the jawbone of an orangutan. The 600-year-old skull fragments and the 500-year-old jaw had been artificially stained to give them the appearance of genuine fossils, and the orangutan's teeth had been filed down to make them look human. Once the forgery was discovered, it rapidly became apparent that the perpetrator or perpetrators hadn't even done an especially convincing job. Although his (assuming there was just one) knowledge of anatomy had been excellent, he had left telltale signs of abrasion on the teeth. Furthermore, two of the molars had been filed down in such a way that they appeared to be slightly out of line with each other. Another tooth, a canine, had apparently erupted only a short time before the specimen died. It too had been filed down, producing counterfeit signs of wear that were inconsistent with its obviously juvenile character.

For more than forty years, distinguished paleontologists remained oblivious to these signs of forgery. When a dentist pointed out the suspicious character of the canine, no one paid any attention. For decades scientists continued to accept a rather ludicrous ape-human combination as genuine.*

In May 1912 Dawson brought Smith Woodward some fragments of a fossil skull that he claimed to have unearthed in a gravel bed on a farm near Piltdown Common in Sussex. Or perhaps he really did unearth them. To this day, it is not known whether Dawson fabricated the "fossil" himself, or whether he dug up a specimen that had been planted by someone else.

Smith Woodward was so impressed by the find that he joined Dawson in a search for additional specimens. The two men were joined by a pair of French priests who were studying at a nearby Jesuit college. One of the priests, Marie-Joseph Pierre Teilhard de

*When Piltdown was discovered, a genuine human ancestor, *Pithecanthropus* (now known as *Homo erectus*), had been known for some twenty years. The Dutch anatomist Eugene Dubois had discovered several fossils of the creature in Java between 1890 and 1892. Unfortunately, the *Pithecanthropus* finds were rather fragmentary—they consisted of part of a jaw, two teeth and a thighbone—and scientists could not reach agreement concerning their significance. Some paleontologists doubted that they even came from the same creature. By the time that it was established that *Pithecanthropus* was the real missing link that paleontologists had been seeking, Piltdown had already been authenticated as a genuine fossil of great evolutionary significance.

Chardin, was, in later years, to gain fame as a paleontologist and as the author of a number of philosophical works on evolution. During the summer of 1912, Smith Woodward, Dawson, Teilhard de Chardin and the latter's priestly colleague carefully excavated the gravel bed. They found additional pieces of the skull, the jaw, and fossil fragments of various extinct mammals, as well as a number of flint artifacts.

When the Piltdown forgery was exposed, it became apparent that the fossil fauna, which included an elephant tooth and a mastodon tooth, had been planted with the Piltdown specimen to make it appear that the skull and jaw were very old. As I have noted previously, good methods for determining the age of a fossil did not exist at the time. However, when a fossil was found in association with fossils of Ice Age mammals, such as the mastodon, this could generally be interpreted as evidence that it too was of great antiquity.

Smith Woodward made a reconstruction of the skull and jaw during the autumn of 1912, and the specimen was exhibited to the members of the Geological Society in December of the same year. The fossil, Smith Woodward told the audience, was much older than the Neanderthal fossils that had been found in Europe. He added that he was inclined to believe that Piltdown was a true human ancestor, while the Neanderthal cave man was an offshoot that had eventually become extinct.

The Piltdown find immediately evoked quite a bit of controversy. A number of different scientists noted that the jaw resembled that of an ape, while the skull looked human. They admitted that the jaw contained teeth that had been worn flat in a human fashion, but insisted that the jaw and skull were fossils of two different creatures.

An answer to these critics was found in 1915, when Dawson came up with a second Piltdown specimen. This one was even more fragmentary than the first. It consisted of nothing more than pieces of a human skull and a molar tooth similar to the ones in the Piltdown jaw.

The "Piltdown II" discovery converted most of the skeptics into believers. As long as only one specimen had been known, it had been possible to maintain that the skull fragments and the jaw had come together by accident. The second discovery, however, seemed to prove beyond any reasonable doubt that the association was a valid one. To maintain that human and ape fossils had been accidentally brought together at two different sites would be to invoke a coincidence that was too improbable to be considered.

Although the Piltdown skull was really of quite recent origin—modern dating methods have established that its original owner died sometime around the fourteenth century A.D.—the scientists who studied it did not hesitate to attribute primitive characteristics to it. Smith Woodward reconstructed the skull in such a way that it seemed to have a cranial capacity of only about 1,000 cubic centimeters, much less than the modern average. Meanwhile, Grafton Elliot Smith, who had taken an interest in the specimen, answered critics who said that the skull was not sufficiently apelike to be associated with the jaw by assuring them that the skull had contained the "most primitive and most simian brain so far recorded."

Gradually, the critics gave way under the weight of expert opinion. The three most eminent British paleontologists of the day, Smith Woodward, Elliot Smith and Arthur Keith, all agreed that the Piltdown fossil was genuine, and furthermore, that it was not a composite. Although Boule in France and a number of scientists in the United States continued to express some doubts, the English soon found themselves in agreement.

National pride may have had something to do with this. Until 1912, most of the fossil evidence of early man had come from continental Europe. Numerous Neanderthal specimens had turned up in France and in Belgium, and a very primitive-looking jaw had been discovered near Heidelberg in Germany. But nothing had been found in the British Isles except a specimen known as "Galley Hill Man" that later turned out to be *Homo sapiens*.

When Piltdown turned up, all this was changed. The French may have had their Neanderthal, but it couldn't compare with Piltdown, which was obviously much older. The first men had turned out to be Englishmen after all!

Another factor that led to the relatively rapid acceptance of the Piltdown fossil was the fact that the specimen conformed quite well to existing theoretical ideas. According to a theory that had been developed by Elliot Smith, it was the brain that has "led the way" in human evolution.

Grafton Elliot Smith maintained that the human brain had acted as a kind of driving force that had brought about the evolution of other modern human characteristics. The brain, in his view, had begun to evolve rapidly before human ancestors had acquired an erect posture, or lost their simian characteristics. Elliot Smith insisted that the association of an apelike jaw with a human brain should not be surprising "to anyone familiar with recent research upon the evolution of man." The Piltdown fossil, in other words, could be viewed as a marvelous confirmation of Elliot Smith's theory.

By the 1940s, Piltdown had become a puzzling anomaly. Fossils of numerous genuine human ancestors had been found, and it had become apparent that Piltdown was not consistent with other findings concerning human evolution. But no one seems to have suspected that the fossil might be a forgery. Piltdown man was regarded, instead, as an enigma that, scientists hoped, would eventually be cleared up. Perhaps he was an offshot from the main evolutionary line who had flourished for a while before going down the path to extinction.

By this time, it had become apparent that human ancestors, such as *Homo erectus*, had possessed brains that were much smaller than the modern average and were certainly smaller than Piltdown's. Nevertheless, they had jaws that were anything but apelike. But no one took up the clue that this observation offered.

In 1947, fluorine tests performed by Kenneth P. Oakley of the British Museum revealed that the Piltdown skull and jaw were of a relatively late date. They were much younger than the animal bones and teeth with which they had been found, and much too young to be in the line of human evolution. But there was still no suspicion of fraud. There was a revival of speculation about the significance of the Piltdown fossil, but nothing more.

Finally, in 1953, it suddenly occurred to the English anthropologist J. S. Weiner that there was only one possible explanation for the association of the skull and jaw in the same geological deposit: they had been deliberately placed in the gravel bed. Weiner communicated his suspicions to his Oxford University colleague Wilfred Le Gros Clark. Weiner and Clark then approached Oakley, who by this time had become head of the Physical Anthropology department of the British Museum of Natural History.

When the three scientists examined the Piltdown specimen together, they saw that the evidences of forgery were obvious. The wear patterns on the teeth looked extremely odd, to say the least. And when the molars were examined under a microscope, evidence of artificial abrasion became visible immediately. In fact, one tooth had been filed down a little too far, and the mistake had been repaired with a plug of a plastic material that looked remarkably like chewing gum.

It was agreed that more extensive tests were warranted. When these were made, it was determined that parts of the fossil had been stained with a paintlike substance (the National Gallery suspected Vandyke Brown paint). New fluorine dating tests showed that the bones were of even more recent origin than Oakley had suspected. In fact, the jaw appeared to be that of a modern ape.

The tests answered every question but one. They did not provide any indication as to who the culprit might have been. As a result, speculation on this subject continues today. Dawson, Smith Woodward, Elliot Smith and Teilhard de Chardin have all been accused by some and defended by others. In 1978, a new suspect turned up. This was W. J. Sollas, who had been Professor of Geology at Oxford at the time of the discovery. But the evidence for his guilt was no more conclusive than that which pointed to the others.

It is not known whether the fraud was the work of one person or the result of a collaboration. The motives for the forgery remain equally mysterious, although there have been quite a number of theories on the subject. Some scientists have speculated that Piltdown was a joke that went too far, that the forger had never really expected that the "fossil" would be accepted as genuine. Others have suggested that professional jealousy was the cause, that Piltdown may have been concocted by a scientist who was trying to damage another's reputation. No one is sure, of course, exactly whose reputation the perpetrator had in mind. It might have been Smith Woodward, who reconstructed the skull, or it could have been Elliot Smith, who had the "brain-first" theory.

And, of course, it is possible that Dawson perpetrated the fraud himself out of a desire for fame. A small piece of evidence that seems to favor this interpretation is the fact that some of the amateur geologists who lived in Dawson's neighborhood spoke privately of the fraudulent nature of the fossil. However, they never made their suspicions public, and it is possible that they really didn't know anything, that their gossip was nothing more than an expression of jealousy. Dawson did not get along well with his amateur colleagues. That seems to be reason enough for them to talk behind his back.

The only thing that is clear is that distinguished members of the British scientific community acted almost as though they wanted to be hoaxed. Shortly after Piltdown was discovered, they waved all suspicions aside and assigned the fossil a place in the ancestry of man.

It would be unjust to ridicule them for doing that. They deserve to be viewed in a more charitable way. After all, they were blundering in the dark, doing the best they could with the few fossils that were available, trying to shed some light on humanity's past. They may have allowed themselves to be influenced by their preconceptions. However, as should be clear by now, there is nothing unusual about that. They were no wiser and no more foolish than the scientists of Darwin's day, or those of our own time.

The examples of Neanderthal and Piltdown should, however, induce us to maintain a certain amount of skepticism when we encounter scientists making far-reaching conclusions about human evolution that are based on evidence that seems not to be very tangible. Those who comment on the subject of human behavior should be subjected to an especially intense kind of scrutiny.

But perhaps a discussion of this topic had best be put off for a while. Before we get into the subject of the discoveries—and possible errors—of the present, we would do well to look at a few more of the mistakes and scientific distortions of the past and take a closer look at some of the theorizing that has been done on the subjects of evolution and human nature between the days of the Neanderthal and Piltdown discoveries and our own time, and at some of the discoveries that have been made as well.

We might do well to begin by looking at a couple of examples of scientific aberration. After all, science—at least when it attempts to deal with so emotion-laden a subject as the nature of man—does not always progress. There have been at least two instances where it has taken great leaps backward.

5

The Aryan Myth and Other Aberrations

When Darwin gave *Origin of Species* the subtitle *The Preservation of Favoured Races in the Struggle for Life*, he was not implying anything about the equality or inequality of human races. He wasn't speaking of race at all, at least not in the modern sense of the term. He was simply using the word *race* as a synonym for *species*.

In Darwin's day, the word *race* had not yet been given a precise meaning. It could mean "species." It could be used in discussions of the white, black and "Mongolian" races. It could be used to designate any people with a common language or ancestry. During the nineteenth century, one could speak of a German or French "race," of the "Hottentot race," even of "the race of Parisians." When the word was used in this last sense, it had a meaning that was not very different from the meanings of the words "nation" and "people."

Perhaps it was only natural that the terminology should have been somewhat confused. During the nineteenth century, biologists did not have the knowledge that would have allowed them to determine just how significant the differences between various races and nationalities were. When Darwin published *Descent of Man* in 1871, biologists had not even reached agreement on whether black and white human beings were members of the same species.

It was this confusion that led to the myth of the Aryan race. At the beginning of the nineteenth century, German philologists were able to prove that there was a similarity between Sanskrit and most European languages. They gave the name *Aryan* to this group of languages, deriving the term from the Sanskrit word *ārya,* which means "noble." Their conclusions are still valid today; one can still speak of "Aryan languages," even though the term *Indo-European* has become more common.

The Aryans were a group of nomadic tribes that spread out in successive waves from southern Russia and Turkestan during the second millennium B.C. At this time, they colonized large sections of India, plundering a flourishing civilization in the Indus Valley, and migrated into Mesopotamia, Asia Minor and Europe. As they did, they spread the language from which numerous modern tongues are derived.

At the beginning of the nineteenth century, little was known about the Aryan migrations. Even less was known about the Aryan way of life. Nevertheless, many scholars of the day jumped to the conclusion that the Aryans had been a "race of conquerors." In their eyes, the fact that English, French, German, Greek, Latin and other tongues were all Aryan languages was proof of this; and so, scholars and lay people alike began to speak of a supposedly superior "Aryan race." Speculation on this subject led, in turn, to discussion of who the Aryan's modern descendants might be.

The results were about what one would expect. German anthropologists said that it was the Germans who were descended from the Aryans, while the French claimed the Aryan heritage for themselves. In England, it was maintained that the true Aryans had been the Saxons. And, since the United States was an English-speaking nation, the latter idea also gained currency there.

Of course, they were all wrong. Whoever the original Aryans were, they certainly intermixed with the inhabitants of the lands to which they migrated. And they were probably a rather diverse collection of tribes to begin with. Most likely, no such thing as a "true" Aryan ever existed.

Unfortunately, the Aryan myth turned out to be just what nineteenth-century Europe was seeking. It seemed to provide the perfect justification for imperialism and for colonial expansion. To be sure, there was nothing new about ideas of national or racial superiority. However, it was only after the discovery of the Aryan family of languages that there seemed to be scientific "proof" that Europeans were descended from a race of conquerors.

Today the myth of Aryan superiority is so closely identified with Adolf Hitler and his National Socialist (Nazi) party that it is easy to forget that the myth gained currency in many other countries. In the United States, for example, it was transformed into a mystique of Anglo-Saxon superiority and was used to justify Anglo-Saxon racism and American imperialism. It influenced political theory as well. During the 1870s and 1880s, American historians theorized that American democratic institutions were related to the primitive institutions of the Aryans. "Freedom" was thought to be a notion that could be traced back to early Saxon tribes; the American capacity for democratic political organization was thought to be part of a racial heritage.

Near the end of the century, social-Darwinist doctrines were combined with the Aryan myth in an attempt to justify militarism and national expansion. It was maintained that the concept of *survival of the fittest* could be applied to "races" as well as to individuals. It was inevitable, the expansionists thought, that the superior races should govern.

During the era of American expansion, such ideas were frequently expressed by Senators Albert J. Beveridge and Henry Cabot Lodge, by Theodore Roosevelt and by Roosevelt's Secretary of State, John Hay. For example, Beveridge spoke these words before the Senate in 1899, when Congressmen and Senators were debating the wisdom of the proposed annexation of the Philippines:

> God has not been preparing the English-speaking and Teutonic peoples for a thousand years for nothing but vain and idle self-admiration. No! He has made us the master organizers of the world to establish system where chaos reigns. . . . He has made us adept in

> government that we may administer government among savages and senile peoples.

Roosevelt too seems to have believed in Aryan superiority. However, he seems to have thought that the Aryan's "great fighting, masterful virtues" could be lost through disuse (a very Spencerian idea, by the way). In "The Strenuous Life," an essay published in 1899, two years before he became President, he exhorted his fellow Americans to disdain "swollen, slothful ease and ignoble peace," and to strive to fulfill their racial destiny. If they did not do so, he warned, then "the bolder and stronger peoples will pass us by, and will win for themselves the domination of the world."

Although Roosevelt and his political associates may have been guilty of harboring racist attitudes, it wouldn't be quite fair to compare them to Adolf Hitler. Not only was Hitler responsible for the deaths of six million Jews and three million Russian prisoners of war, he also practiced genocide against the Gypsies, and he systematically murdered Polish priests and intellectuals so that the Polish "race of slaves" would lack leaders. Many political leaders have made references to the myth of Aryan superiority, but only Hitler used it to justify programs of mass extermination.

Although Hitler combined the Aryan myth with a vehement anti-Semitism, he did not consider the Jews the only inferior race—every non-Aryan people, in his view, was less than fully human; blacks were "half-apes"; Slavs were only fit to be slaves. (This was somewhat inconsistent; the Slavic languages are Aryan too.) According to Hitler, every racially mixed people was inferior; interbreeding "defiled the race" and produced "monstrosities halfway between man and ape." Even Germans of partly Aryan ancestry had no right to live if they were insane, mentally deficient, deformed, or sufferers from any hereditary disease.

Hitler singled out the Jews for attack because he held that they were part of an international "Jewish-Marxist" conspiracy to contaminate Aryan blood, and to subject the German people to "racial disintegration." In *Mein Kampf*, the book that he wrote while serving a prison sentence before his rise to power, he describes the Jews as "black parasites of our nation" who "defile our inexperienced young blond girls." The Jews, he claimed, were not only trying to destroy the German nation through "international high finance" and by Marxist agitation, they were bent also on seducing Aryan women, causing racial degeneration and spreading venereal disease in the process.

It is difficult to view Hitler's paranoid vision with anything resembling calm objectivity. Nevertheless, we ought to make the attempt. When one examines Hitler's racial theories in detail, it becomes apparent that he incorporated the thought of numerous evolutionary theorists into his philosophy. One should make the attempt to understand Hitler's thinking on the subject of race if only because it provides an example of the uses to which ideas about evolution and human nature can be put.

The theory of natural selection, the social-Darwinist concept of the "survival of the fittest," Lombroso's criminal anthropology and Galton's ideas about eugenics all found their way into Hitler's thought. Hitler makes so much use of these concepts that *Mein Kampf* can be read as a kind of perverted evolutionary ideology.

It is true that he says nothing about the past evolution of the Aryan race, and that he does not seem to think that the Aryans will evolve further in the future. However, he appeals to the principle of natural selection over and over again. He combines this idea with mistaken notions about "degeneration."

"Bastardization," Hitler says, always produces inferior individuals. In particular, interbreeding between Aryans and other peoples inevitably leads to racial decline. But fortunately, he goes on, nature has the ability to make certain "corrective decisions with regard to the racial purity of earthly creatures." Since racial hybrids are less fit than individuals of relatively pure ancestry, they tend to be weeded out in the struggle for existence. Thus, as long as the Aryan race does not allow its "blood" to be "defiled" to too great an extent, a natural process of racial regeneration will take place. Furthermore, the process of natural selection will be aided by the inclination of the racially pure to seek one another out as mates. "The results of bastardization spontaneously recede to the background," Hitler says, "as long as a stock of racially pure elements is still present and a further bastardization does not take place."

In Hitler's view, it was the Jews who were trying to promote such bastardization by spreading the idea of equality. If they were successful, he warned, this would lead to the end of Western civilization; of all the human races, the Aryans were the only ones who were capable of creating culture. If their blood became too diluted, then humanity would fall into a dark age from which it would never recover.

Lombroso had characterized criminals as evolutionary throwbacks who were marked by certain physical stigmata. Hitler extended this idea to the entire Jewish "race." The Jews, he said, were a race of degenerates, who combined criminal behavior with certain "lower" physical characteristics.

Surprisingly, Hitler did not claim that the Jews were less intelligent than Aryans. In fact, he seemed to think that intelligence had little to do with Aryan superiority. "The Aryan," he said, "is not greatest in his mental qualities as such, but in the extent of his willingness to put all his abilities in the service of his community." In other words, the superiority of the Aryan consisted of his altruism, his willingness to act in a self-sacrificing manner for the good of his community and the state.

Hitler was no more willing to accept the idea that political equality was desirable than he was to believe in the biological equality of the various human "races." He used social-Darwinist arguments in an attempt to show that democracy would only further the purpose of those who were bent on promoting further racial decline. The parliamentary form of government, Hitler said, had been created by Jews and Marxists, who wanted to "exclude the preeminence of personality in all fields of human life." In an ideal state, he claimed, leadership would descend on those "to whom Nature has given special gifts for this purpose."

Expanding on this idea, he wrote a passage that sounds almost as though it had been taken from a work by Herbert Spencer:

> The selection of these minds, as said before, is primarily accomplished by the hard struggle for existence. Many break and perish, thus showing that they are not destined for the ultimate, and in the end only a few appear to be chosen. In the fields of thought, artistic creation, even, in fact, of economic life, this selective process is still going on today, though, especially in the latter field, it faces a great obstacle.

The "obstacle," of course, was "Jewish-Marxist" democracy.

Galton's theories about eugenics had filtered into Germany long before Hitler attained power. They were promulgated by distinguished German scientists under the ominous-sounding name "race hygiene." When Hitler wrote *Mein Kampf*, he made use of eugenist doctrines in a number of different ways.

Hitler's theory of racial degeneration can be traced back to Galton, who had suggested that the decline of Greek civilization had been caused by interbreeding. But Hitler promoted eugenics with a brutality that Galton would never have dreamed of. In 1939, he put a "euthanasia" program into effect. Every mental institution in Germany was required to fill out a questionnaire identifying each patient and giving the length of time that he had been institutionalized. Those who were considered to be hopelessly insane, retarded or senile were to be put to death. So were all patients who happened to be Jewish. The decree was also to be applied to the physically deformed; according to Lombroso's theories, physical deformity and mental "degeneration" went together.

But even Hitler could not ignore public opinion. The decree was rescinded after German clergymen organized protests against it. Nevertheless, the euthanasia program continued sporadically until the end of the war. It is estimated that some seventy thousand people were executed, among them a large number of malformed children and infants. Interestingly, this is approximately the same number who were sterilized for eugenic reasons in the United States.

Other kinds of eugenic programs too were instituted. In 1936, the *Lebensborn* program was started. Mothers bearing racially pure offspring were to get the best of medical care, whether they happened to be married or not. Racial screening became standard practice in the SS, and SS men were encouraged to have as many children as possible.

Good physical attributes were valued above everything else. In *Mein Kampf*, Hitler had proclaimed that mental qualities were to be considered of secondary importance. "A people of scholars," he had written, "if they are physically degenerate, weak-willed and cowardly pacifists, will not storm the heavens. . . . *A decayed body is not made the least more aesthetic by a brilliant mind*" (italics in the original). Physical health, he insisted, came first, and physical training should be an important part of everyone's education.

When Hitler speaks of perfect physical specimens, we tend to think of the blond, blue-eyed Nordic. But Hitler really gave little emphasis to these particular physical characteristics. When he uses the term "Aryan," he is almost always referring to any German who does not happen to be Jewish.

It was Alfred Rosenberg, the official philosopher of the Nazi movement, who identified the Aryan race with the "Nordic type." It was Rosenberg who insisted that the ideal man as a tall blond with "shining eyes" and a high forehead. Rosenberg suggested that the Nordics had originally come from the lost continent Atlantis. According to his theory, these former inhabitants of Atlantis had provided the ruling caste for every major civilization, including the Greek and Roman ones. Rosenberg claimed that every great Italian of the Renaissance was a blond Nordic who had strayed south by accident. Jesus, he said, had been a Nordic anti-Semite; he certainly couldn't have been a Jew.

Some of the passages in Rosenberg's writings are unintentionally humorous. For example, the claims that the Italian poet Dante Alighieri was really a German named Durante Aldiger. Presumably, Dante changed his name when he traveled south. It was clear, Rosenberg added, that Dante (or Durante) preferred blond women. For that matter, so did Euripides, Aristophanes, and other great men of the past who shared the "Nordic soul."

During the 1920s, many Germans thought that the Nazi movement was doomed because the doctrines promulgated by Hitler and by people like Rosenberg were so easily refuted. They didn't believe that Nazism could triumph; in their eyes, it was based on ideas that were too ridiculous.

But the Nazis did triumph. Hitler assumed power in 1933. By 1939 he had begun a war of conquest, and was attempting to enslave the "inferior" peoples of the world. By the end of World War II, he had all but eradicated the Jews in Germany and in Poland, and had murdered countless other Jews who resided in countries that came under Nazi domination. In retrospect, it is not Hitler who seems ridiculous, but the critics who thought that they could overwhelm him with intellectual arguments.

Ideas about race, human nature and the evolutionary struggle can be used for political ends, whether they happen to be true or not. A Hitler is not required for the accomplishment of this task. As we have seen, such ideas were used to justify racism, imperialism and the inequalities of capitalist society long before there was any such thing as German fascism. Hitler distorted scientific thought to further his own ends. But he was not unique in that respect. Theodore Roosevelt, to cite just one example, was doing something very similar when he appealed to social-Darwinist ideas of biological "struggle" to urge militarism and expansion upon his countrymen.

The thing that made Hitler unique was that he was a mad genius who was able to pursue such ideas obsessively and sway an entire nation when he spoke about them. When Roosevelt spoke of the struggle between races, Americans paid him some attention. But when Hitler weaved together his insane notions about Aryan superiority, degenerate Jews and inferior races, millions of human beings died.

* * *

At approximately the same time that a racist "anthropology" was being developed in fascist Germany, an equally absurd kind of biological "science" was being promulgated within the Soviet Union. Although Soviet "genetics" did not lead to any program of mass genocide—the only people who lost their lives were a few of its scientific opponents—it was made into a state dogma under Soviet premiers Josef Stalin and Nikita Khrushchev. Its proponents proclaimed its ideological virtues, denigrating the true science of genetics as "a tool of United States imperialism." And when, in the end, the defects of Soviet genetics were revealed, recognition of its failures became one of the factors that led to Khrushchev's fall.

In 1948, hundreds of Soviet scientists, including some of that nation's most distinguished biologists, were purged. Although few received prison sentences, a great number were either demoted or dismissed from their positions. The list of crimes of which the victims of the purge were accused was quite a long one. Among other things, they were charged with the following: idealism, reactionary views, Morganism, Weissmannism (Morgan and Weissmann were Western biologists who made significant contributions to genetics), complicity with imperialism and the bourgeois, Mendelism, anti-Michurinism, groveling before the West, sabotage, metaphysics, mechanism, racism, cosmopolitanism, formalism, unproductiveness, anti-Marxism, alienation from practice, and anti-Darwinism.

In reality, they were guilty of one thing: opposition to the theories of the Soviet "peasant scientist" Trofim Denisovich Lysenko.

Lysenko was a Soviet agronomist who had little scientific training. In the 1930s he achieved fame within the Soviet Union by publicizing the success of agricultural practices that he had developed. In reality, these practices were of little value; in many cases they even proved to be detrimental to Soviet agriculture. Nevertheless, Lysenko's influence increased throughout the decade. As it did, he developed a Lamarckian theory of inheritance of acquired characteristics and engaged in disputes over genetic theory with more orthodox Soviet scientists.

In 1938, Lysenko replaced his scientific opponent, the geneticist N. I. Vavilov, as President of the Lenin Academy of Agricultural Sciences. Two years later, Vavilov was arrested, tried, and sent to a Soviet prison, where he was allowed to die of malnutrition. Meanwhile, Lysenko assumed Vavilov's other scientific posts.

In 1948, by order of Stalin, Lysenko's theories were proclaimed official state teachings. Lysenko's critics were purged, scientific journals were subjected to censorship, and textbooks were rewritten to bring them into line with all new official dogma. Lysenko's portrait was hung in all state scientific institutions, and a hymn honoring him was placed in the repertory of the State Chorus.

Lysenko's patron Stalin died in 1953. However, when Nikita Khrushchev assumed power, he saw to it that state support of Lysenko's theories was continued. Lysenkoism continued to dominate Soviet biology until Khrushchev was deposed in 1964 (partly for the failure of his agricultural policies). When Khrushchev fell, Lysenko fell with him, and the traditional scientific theories of genetics were allowed to revive.

Suddenly it again became possible to study biological questions in a scientific manner. For the first time in more than fifteen years, the Soviet government was no longer dictating the correct positions on scientific questions. But of course, by this time, the damage had already been done. Research in genetics had been all but demolished, and Soviet agriculture had suffered setbacks that had cost billions of rubles. The effects of the setbacks would continue to be felt for years to come.

To some observers in the West, it appeared that Lysenko's theories were an outgrowth of Soviet ideology. Others assumed that Lysenko was nothing but a charlatan who had somehow managed to gain a great deal of political influence. As we shall see, neither interpretation was entirely correct, even though there was an element of truth in each of them. Although Lysenko did not base his theories on Marxist dogma, they were used to justify a Marxist outlook. Lysenko seems to have been, not a conscious charlatan, but a sincere, dedicated man who was too ignorant to understand that his theories were of no real value.

In order to understand the factors that were responsible for Lysenko's rise, it is necessary to begin where Lysenko did, and look at the difference between winter and spring wheat. It was Lysenko's experiments on the cultivation of wheat varieties that first brought him before the public eye.

Winter wheat is wheat that is sown in the autumn and harvested during the following summer. Its growth stops during the winter, and starts up again in the spring. The cycle completes itself with the production of seed (this is called *earing*) in the summer. Spring wheat, on the other hand, is wheat that is planted in the spring. Spring wheat completes its cycle of growth in a single season.

The winter and spring varieties are not the same. If winter wheat is planted in the spring, it will not go through the same cycle that is followed by spring wheat. Deprived of a winter chilling period, it will continue to grow without earing. Winter wheat produces higher yields than the spring variety, provided that it manages to survive the winter. Naturally, this is not always the case, especially in the Soviet Union, where winters can be quite severe. Therefore, before one plants it, it is necessary to weigh the advantages and disadvantages. One may obtain a high yield, but there is always the risk that the entire crop will be lost.

During the winter of 1927–28, this is exactly what happened. Some five million hectares of winter wheat perished, chiefly in the Ukraine. Almost immediately, Lysenko set about looking for a solution to the problem. He soon discovered that if winter wheat seed was soaked in water and then chilled, it would ear if it was planted in the spring.

There was really nothing new about this discovery. Lysenko didn't know it, but this technique had been known since the nineteenth century. It had never been put into practice for a very simple reason—it didn't work. The method made it possible to grow winter wheat in the spring, but it didn't increase yields. But

Lysenko wasn't aware of this either. He persuaded his peasant father to plant wheat that had been subjected to this "vernalization" process, as he called it. When the crop came in, Lysenko's expectations seemed to have been confirmed; the vernalized wheat yielded a better harvest than the neighboring spring wheat.

Most likely, the results were due to chance—although, of course, it is possible that Lysenko's father might have put in extra work on the plot out of a desire to make his son's experiment work. Even if this was not the case, nothing had been proved. One trial is never sufficient to establish the usefulness of an agricultural innovation. There are too many factors that can affect the growth of plants. But Lysenko didn't know, or didn't believe, that many more trials would be needed before it could be established that his method worked.

Years later it became apparent that the vernalization method was useless. The yields were not always higher. Sometimes they were lower, and sometimes they were about the same. However, in 1928 no one was aware of this, and Lysenko's "discovery" of vernalization received a great deal of publicity. As a result, the Ukrainian Commissariat of Agriculture ordered large-scale tests of the method. Unfortunately, the Ukrainian officials were a little more enthusiastic than they should have been. They began to publicize the success of the method before the tests had even begun.

This pattern was to be repeated throughout Lysenko's career. Over and over again he developed agricultural techniques that were useless or even detrimental. In nearly every case, the success of the method was publicized before it was tested. After the tests were performed, his success would be publicized again, no matter what the results had been. When a method didn't work, the peasants who had prepared an agricultural plot were blamed, or Lysenko and his followers would resort to selective reporting of data.

For example, when Lysenko developed a new method for growing potatoes, questionnaires were sent to the collective farms that were obliged to try the new process. When the forms were returned, Lysenko sifted through them and reported only the best results.

Lysenko may not have realized that he was doing anything wrong. After all the peasants were frequently not very cooperative. Presumably there were all sorts of factors that could cause an experiment to fail. But, of course, reporting only the best results proved nothing. One would expect that chance factors alone would cause one agricultural plot to produce more than another. Lysenko's experiments, to put it diplomatically, were conducted in an extremely shoddy manner. Nevertheless, his fame grew.

During the 1930s Lysenko developed a theory of genetics that was based on the techniques that he had developed. He believed that vernalization had altered the genetic character of the wheat that had been subjected to the soaking and chilling. He didn't understand that nothing of the sort had taken place, that vernalization only simulated a process that took place naturally when the wheat, in response to the cold of winter, stopped growing.

Lysenko went on to invent other methods that were presumably based on the alteration of hereditary characteristics. By the 1940s he would be claiming that one species could be transformed into another by environmental influences, that wheat could be changed into rye, cabbages into rutabagas, pine trees into firs. One of his disciples went so far as to claim to have changed a rabbit into a chicken.

If such things were possible, then it followed that there could be no such things as genes. In Lysenko's eyes, it was environmental conditions, not a "mythical" genetic material, that determined hereditary characteristics. Lysenko launched bitter attacks on his opponents, the orthodox geneticists. When they attempted to use scientific arguments to show that his theories were incorrect, he replied with political accusations, calling the geneticists "Trotskyite agents of fascism," and associating genetics with Nazi racism.

After Lysenko's theories became state dogma in 1948, all professional scientists and all professors in Soviet universities were required to believe in and to teach Lysenkoist ideas. The study of orthodox genetics was banned, scientific journals were censored, many types of research were halted, and medical researchers were prevented from studying hereditary diseases. All mention of "heredity" had to be removed from scientific works.

And yet the scientific opposition to Lysenko never ceased. When biological journals were no longer allowed to publish articles critical of Lysenkoist theory, such articles began to appear in journals devoted to chemistry, physics and mathematics. This wasn't as dangerous an activity as one might think. Political officials are not known for their willingness to plow through the highly technical and mathematical articles that are published by journals devoted to the physical sciences.

After Khrushchev was deposed in 1964, the Soviet authorities began to actively encourage scientific criticism of Lysenko. By this time, Lysenkoist agricultural practices had been quietly abandoned, even though their success was still publicized. Within the space of a few years, Lysenkoist theorizing had all but disappeared, and Soviet biological science was being brought back into the mainstream.

Obviously, Lysenko's theories were nonsense. However, it is far from obvious why such crackpot ideas should have dominated Soviet biology for two decades, or what motivated the policy of making them part of the state ideology in 1948. Nor is it obvious why the agricultural techniques that Lysenko developed should have brought him such acclaim when they did nothing but impede the progress of Soviet agriculture.

There are no simple answers. Lysenko's ability to propagandize his "achievements" was certainly a factor. So was his peasant background. Soviet authorities were delighted to have an eminent "scientist" rise from a proletarian milieu. In the 1930s, the most noted scientists tended to be individuals who came from upper-class backgrounds and had studied science in prerevolutionary days.

There are two other factors that were at least as important. One had to do with the condition of Soviet agriculture at the time. The second was that Lysenko's theories seemed to imply a view of human nature that was consistent with Soviet ideology.

When Lysenko began his experiments with vernalization in the late 1920s, geneticists and plant breeders had not done much to improve the state of Soviet agriculture. Agricultural practices were still too primitive to make much use of scientific methods, and the peasants tended to resist innovation. They preferred to continue to use methods that were hundreds of years old.

The Bolshevik revolution hadn't helped matters much. In fact, agriculture was in a worse state after the revolution than it had been before. The few landowners who had begun to apply scientific methods in prerevolutionary days had been eliminated. The peasants who had worked for them were happy to have the opportunity to go back to doing things in the old ways.

Many of the Soviet scientists were quite competent. In particular, some of the Soviet geneticists were among the world leaders in their field. However, they could not perform miracles, and miracles were what the Soviet authorities wanted. Believing that the superior socialist system should automatically produce improvements, they could not understand why greater progress had not been made.

At the time, the plant breeders were very much aware of the fact that ten to twelve years were required for the development of new agricultural varieties. When Lysenko promised to produce equal or superior results in three or four years by altering the hereditary characteristics of the varieties with which he worked, he received official support. When the publicity that Lysenko generated seemed to give proof of his success, the support became stronger, and orthodox scientists found themselves victims of state displeasure over their "failures."

By the end of the 1930s, Lysenko had won numerous victories over his opponents among the agricultural scientists. However, it is unlikely that Lysenko would have come to dominate all of Soviet biology if it hadn't been for the fact that his theories meshed so well with Marxist ideology.

According to Marxist theory, Communist revolution automatically brought an end to the social-Darwinist class struggle. Once a socialist state was instituted, human behavior would change rapidly. Under improved political conditions, an egalitarian "classless society" would quickly be forged.

But if it was true that hereditary elements were an important part of human nature, it was not easy to see how this could be the case. Hereditary characteristics, after all, are not changed by environment. Only if environmental factors were most important would human character rapidly improve under the aegis of the ideal Communist state.

But if what Lysenko claimed was true, if there was really no such thing as heredity after all, then matters could be viewed in a different light. A kind of "genetics" that implied that one species could be transformed into another implied that human beings should be able to change as rapidly.

Opposition to German fascism also aided Lysenko. Lysenkoist theories first became influential during the period in which Hitler was consolidating his power. Since Hitler placed great emphasis on genetic differences between individuals and between "races," it was only natural that some of the scientifically unlettered members of Soviet officialdom might begin to wonder whether there was a connection between orthodox genetic theory and racist ideas.

Finally, Lysenkoism was used as a propaganda tool. After World War II, the Soviets began to claim that it was the Western scientists who were promulgating incorrect genetic theories for political reasons. Marxist biology was the only correct biology, they said. Orthodox genetics was a tool of Western imperialism; bourgeois scientists supported it because it served as a justification of the class inequalities that were characteristic of capitalist society. At the time, Stalin was conducting a campaign against "servility toward the West." This provided yet another motivation for denigrating Western biology and enshrining Lysenkoism.

Some authors have maintained that the connection between Lysenkoist theory and Soviet ideology is more apparent than real. They say that Lysenko's activities as a self-publicist and the sorry state of Soviet agriculture are sufficient to explain his rise to power. I find it difficult to accept this point of view. As we have seen, theories about evolution and the inheritance of human character have been used for political ends time and time again. It was no accident that Soviet officials were predisposed to accept Lysenko's ideas about environmentally induced characteristics, just as it is no accident that the capitalist entrepreneurs of the nineteenth century were more than willing to embrace social Darwinism.

✣ 6

The White-Paper Theory Revived

Suppose that it is possible to resurrect a Victorian anthropologist and ask him, "Are human beings naturally monogamous?" What kind of answer would he be likely to give?

I think that it is safe to say that he would reply that all the evidence indicated that they are monogamous. If one objected that there were many peoples who practiced polygamy, it is not likely that he would be disconcerted in the least. "All of the higher cultures are monogamous," he would answer. "If savages practice polygamy, that is a symptom of their lower state of development. In civilized societies, man's better nature has come to the fore. As a result, civilized men find polygamy unthinkable."

But no reputable modern anthropologist would think of answering the question in the same way as our imaginary Victorian. He would not attempt to judge other societies by the standards of his own culture. Today, it is understood that there are no "right" or "wrong" patterns of behavior, no practices that are "more evolved" than others, no traits that make civilized societies intrinsically superior to primitive ones. The modern doctrine of cultural relativism demands that human customs be judged according to the roles they play in their own societies, not by arbitrary standards imposed from the outside. Anthropologists realize that all observed patterns of behavior must be regarded as equally "natural." Ethnocentrism is the worst sin that an anthropologist can commit.

The Victorians, on the other hand, were ethnocentric with a vengeance. The scientists of the day did not hesitate to speak of "miserable, naked savages" who exhibited "fiendish passions." Primitive peoples, in their view, were childlike and cruel; their customs were often disgusting; their art, religion and morality were primitive manifestations of their low evolutionary status.

The Victorians believed that social and biological evolution progressed together. As human beings evolved physically, their savage, barbaric customs were replaced by the civilized virtues. Conversely, when the savages that were encountered did not behave like upper-middle-class members of a European society, this was interpreted to mean that they stood on a lower rung of the evolutionary ladder.

It was commonly believed, for example, that sexual customs had evolved according to a certain set pattern. According to one of the most widely believed theories, universal promiscuity had been the rule during the first, most primitive stage. This promiscuity had been replaced first by cohabitation of brothers and sisters, and then by communal families in which all the men slept with all the women. The next stage was polygamy. The highest was lifelong monogamy in accordance with the Victorian ideal.

Religious systems had supposedly evolved in a similar way. Primitive atheism had been replaced by nature worship or totemism. The next stage was shamanism, and the one after that, idolatry. The highest state was monotheism, which was thought to be associated with the acquisition of high moral precepts. It is not surprising that when the English anthropologist Sir John Lubbock compared the "gross superstitions and ferocious forms of worship" of savages to the higher moral values of the "nobler creeds," few were inclined to disagree.

There was only one problem with these evolutionary schemes. There was practically no evidence to support them. The universal promiscuity that was supposed to represent the first stage in sexual evolution was simply not observed, even among the most primitive peoples. When primitive societies began to be carefully studied, it quickly became apparent that their marriage customs could be quite complex. Furthermore, it seemed that some primitives did practice monogamy, while others who were no more "savage" had various kinds of polygamous systems.

During the latter part of the nineteenth century, anthropologists began, for the first time, to go out into the field and spend time living among primitive peoples. When they did, they began to realize that the reports of primitive life that had been brought back by travelers (who rarely learned to speak the languages of the cultures that they tried to describe) were wholly inaccurate. They began to realize, also, that primitives had virtues, feelings and shortcomings that were not so unlike those observed in Europeans.

The individual who did the most to change anthropological outlooks was the German anthropologist Franz Boas, who began to study Eskimo culture during the late 1880s. Boas soon came to the conclusion that each culture was unique, and that each had to be judged in its own terms. He argued that the behavior of savages provided no indication of hereditary inferiority. On the contrary, complex social factors caused them to behave the way they did. Furthermore, they were neither more nor less rational than Europeans.

Boas drew attention to the lists of "mental qualities" that were supposed to be characteristic of less-evolved people, and he pointed out that it would be equally easy to ascribe these qualities to Europeans. In order to counter the notion that primitives were impulsive and lacking in self-control, Boas concocted the example of a European traveler who had hired natives to accompany him on a journey. The European naturally wanted to reach his destination as quickly as possible. His native employees, on the other hand, did not feel the need to proceed with haste. The European, who was enraged by the delays, concluded that the natives were lazy and impulsive. They were just what he had expected primitive people to be.

But things had an entirely different appearance when viewed from the standpoint of the natives, Boas pointed out. They believed that it was the European who was impulsive and lacking in self-control. Otherwise, he would not have become so irritated by such a trifling cause as the loss of time.

To counter the Victorian idea that white culture was a reflection of the greater reasoning powers of Europeans, Boas suggested that most human institutions, including "civilized" ones, were largely irrational. A culture, he said, was a collection of behavioral forms that had been inherited from the past. Most individuals became unthinking slaves to accepted patterns of thought and behavior, refusing to question them even when they were obviously illogical. Customs persisted because they were customs, even though they might be useless or harmful.

In effect, Boas was replacing Victorian evolutionary doctrine with a modern version of Locke's white-paper theory. Human beings behaved as they did, not because they were acting out hereditary propensities, but because their societies had imposed certain outlooks upon them. Europeans behaved like Europeans because they were brought up to do exactly that. If the same people had grown up in an Eskimo culture, they would act like Eskimos instead.

During the early decades of the twentieth century, Boas's ideas came to be widely accepted. There were a number of reasons for this. One had to do with a scientific reaction against social Darwinism. Sociologists and anthropologists were beginning to realize that social Darwinism wasn't Darwinist at all, that Spencer and his disciples had been wrong when they assumed that social and biological evolution were analogous. If natural selection shaped the physical form of living creatures, it did not necessarily follow that societies evolved in the same way. The wrongheaded use of evolutionary theory to justify various kinds of economic and political doctrines made scientists all the more wary of attempting to apply evolutionary ideas where they were not relevant.

Another cause of the shift in outlook was the decline of instinct theory in psychology. At the beginning of the twentieth century, psychologists tended to believe that there were certain innate human behavior patterns, or *instincts*, that were the analogues of the instincts observed in animals. For example, the American psychologist William James gave the following list of human instincts: climbing, imitation, emulation, rivalry, pugnacity, anger, resentment, sympathy, hunting, fear, appropriation, acquisitiveness, kleptomania, constructiveness, play, curiosity, sociability, shyness, cleanliness, modesty, shame, love, jealousy, parental love. Other psychologists added such items as collecting and hoarding, migration, gregariousness, teasing, tormenting and adornment.

The search for innate behavioral mechanisms had been begun by Darwin himself, who was one of the first experimental psychologists. After he had worked out the details of his theory of natural selection, Darwin became interested in the subject of human emotion. He assiduously set about collecting data. He recorded examples of sounds made by animals as they experienced such things as fear and sexual excitement. He compared these to human vocalizations that were associated with various kinds of emotional response, analyzing loudness, resonance, pitch and timbre. Darwin then conducted similar studies of bodily movements, observing the kinds of motion that were associated with different kinds of "emotive" reactions.

Darwin also persuaded doctors to provide him with data about the emotional behavior of mental patients, and he queried explorers on the subject of expression of emotion in primitive societies. Next, he carried out observations of his own children and made photographs of people while they were experiencing emotion. (He also photographed actors who simulated emotional states.) By the time that he published *The Expression of the Emotions in Man and Animals* in 1872, Darwin had accumulated numerous different kinds of evidence that seemed to show that emotional responses were innate. Some of the examples that he gave were quite striking. For example, there was the case of a girl who had been born blind and deaf who nevertheless laughed and clapped her hands when she was pleased. Obviously, Darwin pointed out, she could not have learned this reaction through imitation.

It would seem that, after all this, it would be difficult to deny that there were innate components of human behavior. Yet this is exactly what the American psychologist John B. Watson did. Watson was not content to argue against the excesses of the psychologists who counted such things as modesty and cleanliness among the human instincts. He denied that human beings had any instincts whatever. According to Watson, all human behavior was the result of conditioning.

Possibly Watson's view was just as extreme as that of the hereditarians. Nevertheless, Watson's work and writings have great significance. If, perhaps as a reaction against the outlooks that were common in his day, he went too far, he did succeed in founding a new school of psychology that was important and influential. If nothing else, he provided an alternative to the theory that most of the elements of human character were innate.

It is interesting to compare Watson's theory of human nature to that expounded by Galton and the eugenists. According to the latter, hereditary qualities were the only things that mattered. Watson, on the other hand, claimed that "there is no such thing as an inheritance of capacity, talent, temperament, mental constitution and characteristics." Given the proper conditioning, a human being could be forced into any mold whatsoever. In his book *Behaviorism*, published in 1925, Watson made the following audacious claim:

> Give me a dozen healthy infants, well-formed, and my own specified world to bring them up in and I'll guarantee to take any one at random and train him to become any kind of specialist I might select—doctor, lawyer, artist, merchant-chief and, yes, even beggar-man and thief, regardless of his talents, penchants, tendencies, abilities, vocations, and race of his ancestors.

Although Watson made mention of "talents, penchants, tendencies, abilities, vocations" he did not believe that such things were innate. He believed that, with the possible exception of certain defects caused by glandular disease or inborn mental deficiency, there were no innate human characteristics. Environment and conditioning were all that was important.

The school of psychology that Watson founded is known as *behaviorism*. Watson's behavioristic theories were only partly motivated by a desire to repudiate the concept of human instincts. Behaviorism was also a reaction against another trend in late-nineteenth- and early-twentieth-century psychology, known as *introspection*.

When Watson studied psychology around the turn of the century, psychology was defined to be the study of consciousness. A psychologist was someone who studied the things that went on in his own mind, and in the minds of other "trained observers." This technique had nothing to do with what is known as "self-analysis" today. The introspectionist psychologists were not concerned with neurosis, with emotion, or even with thinking. A typical introspectionist experiment was one in which a subject sat in front of a screen and looked at certain stimuli, like three dots or the numeral 5. He then had to report what went on in his consciousness when the stimuli were presented. He might find that the numeral 5 evoked a vague feeling of effort in his mind, for example.

In Watson's view, this kind of experimentation was futile. Psychologists should not study consciousness, he said. If they were to have any hope of understanding human behaviors, they would have to look at human behavior. Only by doing that could they obtain results that were meaningful. Watson professed to be unsure that the things called *mind* and *consciousness* even existed. He claimed that psychology had come up with no objective evidence for their reality. In any case, they were concepts that a scientific psychology—behaviorism—could do without. They should be discarded as quickly as possible.

Watson based his theory of human behavior on the experiments that the Russian physiologist Ivan Pavlov had performed on dogs. Around the end of the 1880s Pavlov had become interested in the secretions that were associated with the ingestion of food. He noticed that salivation, a reflex that takes place when food is felt in the mouth, often happened before his experimental dogs were actually fed. Pavlov quickly realized that the stimuli associated with feeding—the sight of the food container, for example, and the smell of the food—were setting off salivation before the dogs had any chance to eat. This gave him the idea of conducting an experiment in which a buzzer was sounded when food was given to an animal. Pavlov found that, after a number of repetitions, salivation took place when the buzzer was sounded alone. Pavlov had succeeded in setting up a *conditioned reflex*.

According to Watson, all human behavior was made up of Pavlovian conditioned reflexes. At birth, Watson said, human beings were alike. The only thing that made them different was the fact that different individuals were conditioned by different stimuli. Although anyone could see that there were numerous differences in bodily structure in human beings, the analogous situation did not hold true for the mind. The human mind was completely a product of environment. And if this implied that there was no such thing as criminal behavior or mental illness, this was not so unreasonable a conclusion. Criminals and the mentally ill were simply people who had been badly conditioned.

Watson's theories had a profound influence upon academic psychology. By the mid-twentieth century, the vast majority of the psychologists who taught in universities were behaviorists.* However, by this time, few of them still believed that Pavlovian conditioning could provide an adequate explanation of human behavior. Watson's behaviorism had been replaced by a more sophisticated theory of conditioning that had first been propounded during the 1930s by the American psychologist B. F. Skinner.

Skinner based his theory on a series of experiments that showed that there was another type of conditioning that seemed to be far more important than the Pavlovian variety. Skinner observed that, if a rat was placed in a cage or a box and was rewarded with food when it happened to press a bar, it would thereafter engage in

*On the other hand, few were Freudians. Freud greatly influenced therapeutic practice, popular attitudes and literature alike. However, his theories never made great inroads into university departments of psychology.

quite a bit of bar-pressing activity whether it was rewarded every time or not.

Skinner noted also that only one *reinforcement* (reward of food) was sufficient to condition the bar-pressing behavior into the rat. The difference between this and Pavlovian conditioning was striking. Pavlov had had to pair a buzzer with the presentation of food many times before a salivation response could be conditioned.

Skinner called the new type of conditioning *operant conditioning* to reflect the fact that it occurred when an experimental animal was able to act upon its environment. By comparison, Pavlovian, or "classical," conditioning was a relatively passive process. Pavlov's dogs had been required to do nothing but gradually learn that the sounding of a buzzer meant that food would soon be forthcoming. Skinner pointed out that, of the two, operant conditioning was far more relevant to the behavior of organisms. Animals, after all, spend most of their time engaging in one kind of activity or another. They don't spend much time waiting around to see what stimuli are going to be paired with one another.

Skinner's experiments showed that operant conditioning could mold behavior in complex ways. For example, it was possible to train a pigeon to peck at a certain spot in its cage in a matter of minutes. One proceeded as follows: First the bird was reinforced with food whenever it happened to turn slightly in the direction of the spot. This would immediately increase the frequency of the behavior; the bird would begin to turn in that direction more often. Next, food was withheld until the bird made a slight movement toward the spot. Once this behavior was established, reinforcement was withheld again. The bird was given food only if it moved even closer. Finally, reinforcement was given only when the bird's beak actually came into contact with the spot. This was sufficient to establish the desired pecking behavior. "Operant conditioning," Skinner commented, "shapes behavior as a sculptor shapes a lump of clay."

Although Skinner's experiments had been performed with rats and pigeons, he did not hesitate to apply his results to human beings as well. He understood well enough that human behavior was more complex. But this only implied that there were more ways to reinforce people. One did not have to reward them with food. Money, kind words, social approval, and control of their physical environment were just a few of the things that they found to be reinforcing.

Like Watson, Skinner felt that "mentalistic" terms like *mind* and *consciousness* had no place in a scientific psychology. He did not deny that consciousness existed, but he did maintain that psychology should deal only with objectively observable units of behavior. The things that went on inside an individual's head were irrelevant because he was the only one who was aware of them.

Today, Skinner admits that natural selection plays a role in shaping behavior. For example, he states that maternal behavior "is a kind of help which is either part of an organism's genetic equipment or which is quickly acquired because of a genetic susceptibility to reinforcement." He also says that "it is part of the genetic endowment called 'human nature' to be reinforced in particular ways by particular things."

However, Skinner's brand of behaviorism is just as deterministic as Watson's was. If human behavior is shaped by conditioning, this may leave some room for the operation of innate propensities. But, even in such a case, there can be no such thing as free will or creativity. It makes no difference whether the conditioning is of the Pavlovian or the operant variety. Where there is conditioning of any kind, there is no free choice. Skinner has commented that "what we call the behavior of the organism is no more free than its digestion, gestation, immunization, or any other physiological process."

Skinner's psychology has been criticized by numerous authors who object to its determinism. The criticisms leveled by humanists who wish to retain the concept of human creativity have been especially intense. Nevertheless, Skinner's most vehement critics have frequently found it necessary to accept some of his ideas.

In one sense, we are all behaviorists today. For example, most of us are no longer willing to believe that criminal behavior is innate. We are more likely to attribute the existence of crime to the effects of the environment. With the exception of those who still persist in believing that black people are genetically inferior, we attribute the poor socioeconomic position of black people to the effects of the ghetto environment. Even those who are most insistent about the necessity of retaining the concept of free will have accepted the idea that environmental influences have some conditioning effect.

Behaviorist outlooks have had a significant influence on anthropology. This is only to be expected. After all, psychology and cultural anthropology are profoundly related. One deals with individual behavior, the other with the behavior of groups of human beings. It is not surprising that there should exist anthropological studies that throw light on the ways in which human nature is conditioned by human societies.

One of the best-known of these studies was conducted by the American anthropologist Margaret Mead and is described in her book *Sex and Temperament in Three Primitive Societies*, published in 1935. Although Mead was not, strictly speaking, a behaviorist—she was influenced by Freudian outlooks at least as much—her work is not in the least inconsistent with behaviorist viewpoints. In fact, she was able to show how some of the cultural conditioning to which human beings are subjected takes place.

During the years 1931 to 1933, Mead studied three primitive societies in New Guinea: the Arapesh, the Mundugumor and the Tchambuli. She found that the Arapesh were a gentle, unaggressive, cooperative people. Interactions among the Mundugumor, on the other hand, were characterized by suspicion and hostility. In each tribe, males and females exhibited similar traits. In Arapesh society, men were not more aggressive or more violent than the women. Similarly, Mundugumor women were just as violent, competitive and jealous as their men.

In the third society, that of the Tchambuli, the roles of men and women were practically the reverse of what they were in Western culture. Not only were the women believed to be more highly sexed than the men, they also tended to be more aggressive and competitive. It was the women who fished and made trade goods in order to provide for their families, while the men were more concerned with personal adornment and with artistic pursuits.

Mead did not restrict herself to making observations of adult behavior patterns. In order to learn how this behavior was molded, she paid a great deal of attention to child-rearing practices. Not surprisingly, she found that children were trained for their adult roles.

The conditioning began at birth. Arapesh infants were coddled; they were suckled whenever they cried; and they were allowed to sleep in close contact with their mothers' bodies. As children grew older, they were given a great deal of attention. There was no insistence upon the need to grow up rapidly, and children were not encouraged to engage in competitive games. Fighting was discouraged, and children were taught not to exhibit anger.

In most cases, the conditioning was so effective that the Arapesh shied away from competition for dominant roles. Assuming the role of a "big man" within the Arapesh society was regarded as a distasteful and onerous task. The big men were those who, for the sake of the community, organized feasts, raised pigs, undertook journeys and made trade contacts with other tribes. In most cases, the big men did not relish their socially dominant positions. On the contrary, they looked forward to the time when they could pass their responsibilities on to their sons and gracefully retire. The Arapesh big men, unlike the big men in our society, were not individuals who were consumed with ambition, who had thrust themselves to the forefront. The role had to be imposed upon young men against their will.

In the Mundugumor society, infants experienced little physical contact with their mothers. They were kept in stiff baskets which pinioned their arms to their sides. When they cried they were ignored. It was only when the crying became persistent that their mothers reluctantly suckled them. Even then a baby would not be fondled; it would be encouraged to quickly absorb the food that it needed so that it could be put back in its basket without any further nuisance. Sometimes the babies choked from attempting to swallow too fast. When this happened, the mothers did not show concern; they became angry with their infants instead.

Mundugumor children were taught from birth that life was a struggle. As they grew older, they were encouraged to engage in fiercely competitive games and to bully, threaten and tease children who were smaller. Adults too subjected children to bullying and a certain amount of knocking about. The children, on the other hand, were given license to insult their parents and attempt to humiliate older people.

The aggressive patterns of behavior continued into adult life. Relations within a family were not particularly tender. They were likely to be even more hostile than those with outsiders. Fathers and sons were often one another's worst enemies, and for good reason. If a Mundugumor man wanted to obtain a wife, he was expected to give a woman in return, usually his sister. But a man's father could use these women—his daughters—to obtain additional wives for himself. Since prestige and power within the society was measured by the number of wives that a man possessed, there was constant conflict between fathers and their male offspring.

Of course, the Mundugumor did not realize that they were conditioning their children to think, behave and feel in such ways. In their eyes, hostile and suspicious behavior were only expressions of "human nature." Similarly, the Arapesh thought that it was natural for human beings to be gentle and noncompetitive. They were so committed to this view that they did not know how to deal with the few deviants within their society who turned out to be aggressive or violent.

In the Tchambuli tribe, it was believed that it was human nature for women to be dominant and sexually aggressive, while men were naturally passive, responsive, and interested in children. And the Tchambuli had no idea that they were conditioning boys and girls to accept different roles.

In a sense, Mead's results were almost too good to be true. In one society, both men and women were aggressive. In a second, members of both sexes exhibited what we in the West would call "feminine" traits. And in a third society, the respective male and female roles were roughly the opposite of what they have traditionally been in our own.

But Mead had not been looking for societies that exhibited these particular characteristics. She simply reported what she found. Under other circumstances she might have wound up studying different tribes and writing a different kind of book. Certainly, chance played a role in her selection of the Arapesh, Mundugumor and Tchambuli tribes as objects of study. However, that does not affect the validity of her results.

In the preface to the 1950 edition of *Sex and Temperament*, Mead commented that it had been her "most misunderstood book." Most of those who had read it had seen it as an anthropological treatise on the conditioning of behavior. To those readers, it seemed that Mead was saying that innate factors played no role in the development of human personality.

Yet Mead had discussed innate temperamental differences throughout the book. She had stressed that, although each society tried to mold behavior to a certain cultural norm, each of the three also contained deviant individuals upon whom the conditioning seemed to have little effect. There were violent individuals among the Arapesh, and relatively unaggressive, gentle people could be found among the Mundugumor. Similarly, there were Tchambuli men and women who could not adjust to the roles prescribed for their respective sexes.

Mead felt that the existence of deviant individuals in these societies was evidence of innate temperamental differences between human beings. If human character was perfectly malleable, she said, then there would not be any instances where the conditioning failed to "take." She suggested that there were genetic components in human behavior, and that an entire range of innate temperaments was possible. Mead suggested, also, that each society had selected one of these possible human character types and had imposed it as a cultural norm.

In other words, perhaps some human beings were naturally

aggressive, while others were naturally gentle. The Arapesh had tried to make everyone conform to the latter type, while the Mundugumor had selected the former as their ideal. Since everyone could not be fitted into the same mold, each society contained individuals who could not adjust.

This is an interesting hypothesis. However, one must realize that the evidence that Mead gave in support of it is somewhat less than compelling. She demonstrated conclusively that societies managed to form the character of the majority of the individuals. The idea that deviance was caused by innate temperamental differences must be viewed as somewhat more tentative. After all, it is possible to make the alternative hypothesis that some individuals are deviant because they have been badly conditioned. One is not forced to agree that the deviants must have been temperamentally different from birth.

So, perhaps Mead's readers cannot be blamed for paying most attention to the environmentalist implications of her book. However, one must not conclude that she was wrong about innate temperaments. In the last few years, studies of animal behavior have produced some very interesting results, which suggest that she may have been right after all. Although studies of animals can tell us nothing conclusive about human beings, they can be suggestive. One study in particular suggests very strongly that Mead may have been on the right track.

It seems that wolf cubs in a given litter will exhibit different temperaments. Tests indicate that they possess varying degrees of aggressiveness. The red fox, on the other hand, produces litters in which the cubs are on a par with one another as far as aggressiveness is concerned.

One would think that these results tell us very little. After all, they seem to say nothing but that wolves are temperamentally different while foxes are not. But this is not the case, for there is a very important difference between wolves and red foxes.

The foxes are solitary animals, while the wolves run in packs. The differences in aggressiveness may help the wolves to form dominance hierarchies; the more aggressive cubs are the ones that are most likely to grow up to become pack leaders.

Like wolves, human beings are social animals. Humans have dominance hierarchies too. If it is true that some people are more aggressive and competitive from birth, they will be the ones most likely to rise to high positions in human society. It may be that in both wolves and human beings evolution has taken pains to ensure that there are individuals who will be natural leaders.

Such a conclusion must, of course, be considered to be very tentative. It is worth emphasizing again that one cannot use observations of the behavior of wolf cubs to draw irrefutable conclusions about people. Furthermore, human character is much more influenced by learning than the character of any animal. Only humans have culture, and only humans possess a brain that makes them almost infinitely adaptable.

But one thing is clear. There are almost certainly innate components of human behavior. If behaviorists like B. F. Skinner find it necessary to admit that this is the case, and if studies that, like Mead's, emphasize environment to such a degree nevertheless turn up evidence of innate temperamental differences, it is hard to doubt that biology plays some role in making us what we are.

7

Human Aggression

In 1953, in an essay entitled "The Predatory Transition from Ape to Man," the South African anatomist Raymond Dart painted a grisly picture of human ancestors. According to Dart, they were "carnivorous creatures that seized living quarries by violence, battered them to death, tore apart their broken bodies, dismembered them limb from limb, slaking their ravenous thirst with the hot blood of victims and greedily devouring living writhing flesh."

It was the "crude, omnivorous, cannibalistic bone-club-wielding, jaw-bone-cleaving Samsonian phase of human emergence" of our prehuman ancestors, Dart added, that was the source of aggression in modern human beings. "The loathsome cruelty of mankind to man," Dart said, "forms one of his inescapable characteristics and differentiative features; it is explicable only in terms of his carnivorous and cannibalistic origin."

In 1961, Dart's "killer ape" theory was popularized by the American dramatist Robert Ardrey in his book *African Genesis*. "Man is a predator whose natural instinct is to kill with a weapon," Ardrey told his readers. It was the use of weapons, Ardrey said, that had led to man's bipedal gait and to his large and complex brain. The use of weapons had "perfected the specialized human anatomy, and demanded complex nervous coordination never before experienced in the animal world."

The quality that set man apart from the rest of the species that inhabited this planet, Ardrey said, was his fascination with killing.

"The superior weapon," he maintained, "has been the central human dream." Indeed, Ardrey went so far as to suggest that the only hope for the human species might lie in the possibility that man might evolve into a creature that was somewhat less aggressive after his numbers had been ravaged by the inevitable nuclear holocaust.

Ardrey and Dart were very much aware that many anthropologists stressed the cooperative, rather than the aggressive, side of human nature. But this was an error, they said. Scientists believed in the "romantic fallacy of the innocence of man" because they wanted to believe in it. They attempted to deemphasize the pathological aggressiveness that was characteristic of human nature, because they were unwilling to face the true facts. if there was to be any hope for human survival, Ardrey insisted, then it would be necessary to give up these "heart-warming but obsolete assumptions." It would be necessary to view man as he really was, no matter how unpleasant the picture might be.

In 1963, two years after the publication of *African Genesis*, a book by the Austrian ethnologist Konrad Lorenz was published in German under the title *Das Sogenannte Böse* (literally, "The So-called Evil"). Three years later this book appeared in English as *On Aggression*. Like *African Genesis*, it became a best-seller at once.

Although Lorenz made no mention of Ardrey or Dart, he concurred in their conclusion that man was unique in that he had a "killer instinct" that was not found in any other animal. In human beings, aggression took on a character entirely different from that of aggression anywhere else in the animal world.

Lorenz began by pointing out that many animal species had aggressive instincts. These instincts, he said, had evolved because they performed a number of beneficial functions. The most important of these was defense of territory. Many animals set up territories of one kind or another and defended them against other members of their species. Since only a limited number of territories were available, this had the effect of spacing the animals out, preventing overpopulation.

Although many animals were aggressive, Lorenz went on, they rarely killed conspecifics. In fact any animal that did would labor under a severe evolutionary disadvantage. If wolves habitually killed other wolves, for example, they would rapidly become extinct.

In some species, Lorenz observed, it wasn't necessary for evolution to find ways of limiting aggressive behavior. For example, rabbits didn't do much damage to one another when they fought. In any case, the weaker animal could always run away.

But some animals—wolves, for example—were equipped by nature with powerful weapons. The wolf's strong jaws and sharp teeth would enable it to dispatch weaker members of its species in an instant if nature had not provided a mechanism that would prevent this kind of killing.

The mechanism, Lorenz pointed out, was quite a surprising one. When wolves fought, they generally snapped at each other without doing much damage. Eventually, the weaker animal would tire and perhaps even stumble. What happened next was just what one would expect: the stronger animal would leap in for the kill.

At this point, something very astonishing happened. The defeated animal did not try to defend himself against the onslaught. Instead, he presented the most vulnerable part of his body—his throat—to the victor. Under such circumstances, it would be easy for the stronger animal to sever the jugular vein of the loser in an instant. But this never happened. He would snap at the throat of the other animal, somehow unable to bite. Something—some innate inhibition—prevented him from killing.

Lorenz used this example to point out that animals that were capable of killing one another seemed to have evolved inhibitions against killing any conspecifics who exhibited an appropriate gesture of submission. Wolves presented their throats to the victor. Dogs rolled over on their backs, exposing their vulnerable underbellies. Baboons turned around and presented their hindquarters, imitating the sexually receptive posture of a female. Whenever such a gesture of submission was made, the stronger animal immediately ceased all aggression. Even in those cases where he wanted to continue the conflict, he seemed unable to do so.

Unlike Dart and Ardrey, Lorenz did not claim that human beings had an aggressive drive that was stronger than that of other animals. In his view, the thing that distinguished man from other animals was the fact that he lacked inhibitions against killing members of his own species, while possessing the weapons that made it easy for him to do so. It was this fact which made him the most dangerous creature of any that inhabited the planet.

Many animals lack killing inhibitions. They have failed to evolve in numerous species for the simple reason that they never were needed. Rabbits are just one of many possible examples. This category also includes the ancestors of man; because they did not possess sharp claws, long canine teeth, or other natural weapons, fighting was unlikely to cause serious injury. And, as in the case of rabbits, the loser could always run away.

But all this changed when man's ancestors began to use tools and weapons. Although they still couldn't injure one another severely with their fists, or with their relatively weak teeth, they found killing very easy once they had such things as clubs and axes.

Nothing could be more obvious than the fact that human beings do not club one another to death every time they become angry. Obviously some inhibitions have developed. But, according to Lorenz, this really hasn't solved the fundamental problem. "If moral responsibility and unwillingness to kill have indubitably increased," he said, "the ease and emotional impunity of killing have increased at the same rate." This evolved "moral responsibility" did not inhibit man from waging war, because the "deep, emotional layers of our personality do not register the fact that the crooking of the forefinger to release a shot tears at the entrails of another man."

Lorenz's theory can be summed up as follows: Like all animals, man has an innate aggressive drive. Ever since weapons were developed by our prehuman ancestors, man has had to learn to refrain from using those weapons on members of his own group. But he still lacks the strong inhibitions against killing that animals like the wolf have. As a result, his aggressive instincts can suddenly explode.

Lorenz believed that aggression was an innate drive that had to find an outlet one way or another. In his view, the fact that man could understand the effects of his actions often caused him to refrain from injuring his fellows. But, as a result, the aggressive drive remained bottled up. Sometimes it was released in war. In prehistoric times, Lorenz said, men battled "hostile neighboring hordes." In the modern world, they were threatening one another with nuclear destruction.

More than two decades have passed since Dart, Ardrey and Lorenz began to publish their ideas about the nature of human aggression. During that time a number of scientific studies have cast doubt on the validity of their conceptions. It has become apparent that Dart misinterpreted some of the fossil evidence upon which he and Ardrey based their killer-ape hypothesis. And Lorenz's assumption that human beings are the only animals that habitually kill members of their own species has turned out to be incorrect. Studies of animals in their natural habitats have revealed that many animals kill conspecifics quite frequently. The list includes gulls, langurs, lions, hippos, hyenas, macaques, elephants and chimpanzees. It can only grow longer in the future. Compared to some of these species, human beings are relative pacifists. The number of lethal assaults carried out by humans is relatively small, even when the deaths that have resulted from our periodic wars are averaged in. Lions are more likely to die by "murder" than are inhabitants of big-city ghettos.

It would be absurd to deny that aggression is an important human trait. Our propensity for waging war is just as much a problem as it was when Ardrey and Lorenz wrote their books in the early 1960s. However, it is by no means clear that human beings possess an innate aggressive drive. In fact, there is currently considerable controversy on this point.

According to many anthropologists, aggressive behavior in human beings is learned. It is cultural conditioning, they say, that causes us to be aggressive or gentle. "Everything human beings do as human beings they have had to learn from other human beings," says anthropologist Ashley Montagu. Behaviorist psychologists and others who emphasize the role of learning (psychologists often use the terms "learning" and "conditioning" synonymously) concur.

On the other hand, ethologists—scientists who study animal behavior—tend to agree with Lorenz that aggression is innate. They point to innate patterns of aggression in animals and suggest that there is every reason to believe that similar patterns exist in human beings. For example, the German ethologist Irenäus Eibl-Eibesfeldt insists that an innate predisposition for waging war is programmed into the human nervous system. Although he admits that war is a cultural institution, he maintains that it has biological roots in two inborn human characteristics: our distrust of strangers and our readiness for aggressive action.

It would be tempting, at this point, to jump right in and discuss the current controversy in greater detail. However, I suspect that it might be best to backtrack a little and describe some of the discoveries that have been made concerning human evolution first. After all, if there are any innate behavior patterns in human beings, they were molded by natural selection. If we are to have any hope of understanding what human nature is, then it is necessary to know something about our evolutionary past. Perhaps there is no better place to begin than with the discovery of the australopithecines, the creatures that Dart and Ardrey characterized as killer apes.

In 1924, Josephine Salmons, a student at South Africa's Witwatersrand University, noticed what she took to be a fossil baboon skull on the mantelpiece of a friend's living room. When she asked him about it, he told her that there were no monkey or ape fossils in South Africa; he didn't know what it was. So Salmons told her anatomy professor, Raymond Dart, of the fossil. He asked her to borrow the specimen and to bring it to him for examination.

When Dart examined the fossil, he found that it was indeed a baboon skull, possibly of a new species. After learning that the fossil had turned up during lime-quarrying operations near a place called Taung, Dart began to wonder if any other interesting fossils might turn up there. He enlisted the aid of R. B. Young, a colleague in the university's geology department. Young, who planned to visit the Taung quarry, agreed to ask the quarry owners to preserve any fossils that they might discover.

There are two different accounts of what happened next. According to Young, he arrived at the quarry just after blasting had taken place and discovered that a large piece of rock that had been split in two had a fossil embedded in it. Young says that he packed the find and handed it to Dart after returning to Johannesburg.

In his book *Adventures with the Missing Link,* Dart told a somewhat different story. According to Dart, two boxes of rocks were shipped to him from the quarry. Going through one of the boxes, he came upon an *endocranial cast* (a cast of the inside of a skull formed when limestone fills the brain cavity). At first, Dart supposed that it must be the cast of the inside of the skull of a baboon. But, upon examining it more closely, he discovered that it was too large to fit inside any baboon skull he had ever seen.

Thinking that he might have happened upon a brain cast of previously unknown species of ape, Dart rummaged around in the box for a piece of fossil skull that would fit it. He soon found a piece that fit the cast perfectly.

But all that he could make out when he looked at the fossil skull was the inside of the head. The face of the fossil was covered with an encrustation of breccia, a cementlike substance formed of a mixture of limestone with sand and gravel. Dart spent the next seventy-three days chipping away at the breccia with one of his wife's knitting needles. When he had finished, he saw that his suspicions had been confirmed. The fossil was definitely not that of a baboon. It seemed to be the skull of a young ape that had been perhaps five or six years old when it died.

But the specimen could not be a gorilla or a chimpanzee. It lacked the large canine teeth that modern apes possess. In some respects, it seemed more human than apelike; it lacked the eyebrow ridges that are found in ape skulls, and the jaw did not jut forward. Furthermore, the skull had another very peculiar characteristic. The *foramen magnum*—the hole in the skull through which the spinal nerves pass into the brain—was at the bottom of the skull, rather than toward the rear, as it always is in nonhuman primates.

This suggested to Dart that this creature must have walked upright. In man, the head is balanced on top of the spinal column, and the foramen magnum is positioned differently from that of apes, which go around on all fours (apes have a mode of locomotion which is called knuckle walking; although chimpanzees and gorillas are capable of standing and walking erect, they aren't very good at it).

Although the specimen had possessed a brain that was no larger than that of a gorilla, Dart was convinced that he had discovered an erect-walking human ancestor. Naming the specimen *Australopithecus africanus* ("southern ape of Africa"), Dart submitted a paper on it to the British scientific periodical *Nature*.

But Dart's claim that he had discovered a new "missing link" only aroused skepticism in the scientific community. It was suggested that the fossil wasn't that of a manlike creature at all. Most likely, some British anthropologists said, it was nothing more than a previously unknown species of chimpanzee or gorilla.

There were a number of different reasons for this reaction on the part of the scientific establishment. Some of them may not have had anything to do with the merits of Dart's claim. It seems that Dart's discovery had received quite a bit of newspaper publicity. Headlines about a "missing link" that was "5 millions years old" (actually, Dart had estimated the age at about one million) aroused skepticism among scientists before the paper in *Nature* was even published. Also, Dart seems not to have made it easy for British anthropologists to obtain casts of the specimen so that they could study it and draw their own conclusions. Finally, the fossil skull was that of a child; it was conceivable that an adult specimen would be more apelike in appearance. This was not an unreasonable objection. The skull of an infant chimpanzee, for example, seems very "human" in appearance.

However, none of these considerations really explain why Dart's claims should have been dismissed so unceremoniously. One would think that the British anthropologists would simply have withheld judgment. For some reason, they appeared just a little too anxious to conclude that the specimen was nothing but an ape.

One cannot help but suspect that *Australopithecus africanus* was denied a status as a possible human ancestor, because it would have upset theories that were current at the time. At the time of Dart's discovery the theory that the human brain had evolved first and that bipedal locomotion and other modern characteristics had come later, was still very influential. The "discovery" of Piltdown man had apparently confirmed this theory in a marvelous way. Accepting Dart's idea that a creature with a brain no larger than that of an ape was capable of walking with an erect posture would have upset everything.

It wasn't until the late 1940s that *Australopithecus africanus* was accepted as a possible human ancestor. By this time, numerous specimens of *Homo erectus* had been found in China and elsewhere, and it had become apparent that Piltdown (which had not yet been exposed as a forgery) could not possibly be ancestral to modern man. In 1946, the Oxford anatomist Wilfred le Gros Clark found that there were good anatomical reasons for supposing that the *Australopithecus* fossils (by this time, many more had been discovered) were the remains of human ancestors. In 1947, Sir Arthur Keith, who had originally been one of Dart's harshest critics, wrote that he was forced to concur. "I am now convinced that Professor Dart was right and I was wrong," he conceded.

Today, paleontologists once again find themselves in disagreement on the question of whether *Australopithecus africanus* is really a human ancestor. Some of them, most notably the Kenyan paleontologist Richard Leakey, maintaint that *africanus* was a cousin of man, not an ancestor. That is, he was an evolutionary offshoot which eventually became extinct. The American anthropologist Donald Johanson believes that a species which he had named *Australopithecus afarensis* was the ancestor both of *africanus* and of *Homo*. Some scientists believe that both *africanus* and *afarensis* belong in the human line. Others maintain that neither does. And there are yet others who refuse to recognize the species *Australopithecus afarensis* at all.

Therefore, before we go on to a consideration of the evidence that led Dart to form his killer-ape theory, it might be best to give a brief summary of the facts concerning human evolution. There are facts that can be stated. Even though anthropologists find themselves in disagreement on a number of points, there are some matters on which they all agree.

At some unknown date millions of years ago, the apes diverged into two distinct evolutionary lines, the family *Pongidae* and the family *Hominidae*. Members of the first family are generally referred to as the "pongids," and those of the latter are known colloquially as the "hominids." The former term is applied to anything that is more apelike than manlike. Modern examples of pongids are the chimpanzees, the gorilla, the orangutan and the gibbon. The australopithecines, *Homo erectus* and *Homo sapiens*, on the other hand, are all hominids.

The genus name *Australopithecus* is given to a number of different fossil species, of which *Australopithecus africanus* is just one. Modern man may or may not be a descendant of *africanus*. But he is certainly descended from some species in the genus.

There is some controversy concerning the date at which the australopithecines evolved. However, it is generally agreed that they appeared somewhere between five and nine million years ago. As early as 3.5 million years ago, they were fully bipedal. However, at this time, their brains had not yet grown any larger than those of the apes.

Two million years ago, there were at least two kinds of australopithecines, big ones and little ones. The big ones are further divided into two species, *Australopithecus robustus* and *Australopithecus boisei;* it is agreed that they are not human ancestors. The little ones, sometimes referred to as "gracile australopithecines," are Dart's *Australopithecus africanus*.

Homo erectus, the immediate ancestor of *Homo sapiens*, emerged about 1.5 million years ago. *Erectus* had a brain that was about twice the size of that of the australopithecines and about two thirds as large as that of modern man. He evolved into *Homo sapiens* about 200,000 years ago.

Homo erectus's predecessor is called *Homo habilis*. It is the status of *habilis* that has caused all the controversy. *Homo habilis* may be a descendant of *Australopithecus africanus*, or he may have co-existed with *africanus*. It is assumed that *habilis* must have evolved from some species of australopithecine, but no one is really sure which one. In a way, the controversy resembles that which still goes on concerning Neanderthal man. As we saw in a previous chapter, no one is quite sure whether Neanderthal was an ancestor of modern man, or only a cousin.

I realize that reading through these lists of Latin names can make one feel somewhat dazed. At least, that happens to me sometimes. But perhaps it wouldn't hurt if I summarized it all again, this time in the style of the Biblical "begats."

Between 5 and 9 million years ago, the australopithecines evolved, and diverged into several species. Two million years ago, one of the species of australopithecines begat *Homo habilis*. One and a half million years ago, *Homo habilis* begat *Homo erectus*. Two hundred thousand years ago, *Homo erectus* begat *Homo sapiens*.

A table will help to complete the picture:

TABLE 1

Species	*Cranial Capacity in Cubic Centimeters*
Modern chimpanzees	350 to 400
Australopithecus africanus	430 to 480
Homo habilis	600 to 800
Homo erectus	750 to 1,250
Homo sapiens	1,000 to 2,000

A few other details are worth mentioning: *Australopithecus africanus* weighed 45 to 70 pounds, and was 3½ to 4½ feet tall. *Homo erectus* was 5 to 5½ feet tall, and weighed about 120 pounds. And of course *Homo habilis* was somewhere in between. This is one of the factors that has caused most authorities to view him as a transitional form.

Dart made a number of observations concerning the *Austra-*

lopithecus africanus fossils found in South Africa and drew some very astonishing conclusions about the australopithecine life style and about aggression in modern man. Now, Dart believed that *africanus* was a human ancestor and, as we have seen, this assumption may be incorrect. However, if it is incorrect, this does not destroy the validity of Dart's argument. After all, human beings are descended from australopithecines of some kind, if not from *africanus*, then from one of his near ancestors or close relatives. Any conclusions that can be drawn from knowledge about the evolution of *africanus* may very well tell us something about ourselves.

Most of the South African *Australopithecus africanus* fossils were found in association with the fossil bones of numerous other animals, including antelopes, wild pigs, rhinoceroses, leopards, giraffes, monkeys, horses, tortoises, hares, porcupines, hyenas, baboons and jackals. In Dart's view, these fossils constituted evidence that *africanus* had been a hunter and, furthermore, that he had hunted with weapons. A creature as small as *africanus* could hardly have killed these animals with his bare hands.

It was possible, of course, that *africanus* had been a scavenger, that he had dragged carcasses back to the caves after the animals had been killed and partly eaten by leopards and other carnivores. There was nothing very implausible about this. Even today some primitive peoples scavenge the prey of the big cats. However, Dart thought that he had evidence indicating that the australopithecines had hunted.

It seemed that numerous fossil baboon skulls had been discovered in the same deposits in which australopithecine fossils had been found. Furthermore, many of the skulls were fractured. Suspecting that the baboons had been the victims of intentional violence, Dart enlisted the aid of R. H. Mackintosh, head of the department of forensic medicine at the University of Witwatersrand. Mackintosh concurred in the diagnosis of death by violence. Furthermore, out of some forty-two baboons, twenty-seven seemed to have received blows from the front; only six had been struck from the rear. Of the remaining nine, seven had been struck on the left side (that is from the attacker's right). To Dart, this indicated that *africanus* had been a hunter, and that he had been right-handed.

There seemed to be only one problem with the theory. At the time, no tool or weapon had been found in association with any of the fossils. But Dart thought that he had an answer to this objection. *Australopithecus africanus* had used bones as weapons, specifically the *humerus*, or upper-foreleg bone, of the antelopes that he had killed. These humerus bones, Dart noted, occurred in the caves in great profusion.

Dart did not stop there. There was also, he said, convincing evidence that *africanus* had murdered members of his own species; he had apparently not been content to slaughter only animals. He noted that many of the australopithecine skulls exhibited fractures that were similar to those which had been observed in the baboons. Furthermore, one of the skulls contained two holes that were a little more than an inch apart. No carnivore, Dart said, had canines that were set so close together. The holes must have been made by blows with a weapon.

Although Dart's theories gained a certain amount of popular acceptance, many scientists remained skeptical. In 1972, C. K. Brain of South Africa's Transvaal Museum pointed out that there really wasn't any evidence that the australopithecines had gone around bashing in one another's heads. He observed that the damage attributed to blows with weapons could have been caused by postfossilization effects. When a fossil skull remains in contact with a stone for long periods of time, fractures can result. Buried skulls can also be damaged or deformed by pressure. Since the fossils had remained in the caves for two million years before they were found, there had been ample time for such damage to occur.

Brain also pointed out that Dart had been mistaken when he claimed that the two holes that had been found in one skull could not have been made by a carnivore. Brain found that there was a fossil leopard with lower canine teeth that were precisely the right distance apart. Brain admitted that there was no way to prove that the holes had been made by a leopard that was dragging an australopithecine to its feeding place. Nevertheless, it seemed the most likely explanation.

Brain also performed a series of experiments in which he studied the effects of weathering and the actions of modern scavengers on animal bone collections. After making observations over a period of years, he concluded that the scavenging habits of carnivores and the different resistance to weathering exhibited by various types of bones produced collections that were practically identical to the collections found in the South African caves. The fossils with which *Australopithecus africanus* was found were no indication that he was a hunter after all. The evidence, Brain concluded, seemed to indicate that *africanus* was not a hunter, but rather the hunted.

The australopithecines did use tools, although they probably didn't manufacture them. Objects that look like stone implements have been turned up at fossil sites. And the australopithecines may very well have made some use of weapons. Chimpanzees sometimes throw rocks at baboons, and if an experimenter presents them with a stuffed leopard, they will attack it with sticks. If apes are capable of such behavior, there is no reason to think that the fully bipedal australopithecines did less.

However, it is apparent that Dart was mistaken when he concluded that the australopithecines made extensive use of bone clubs. This particular idea seems to have sprung from wishful thinking. Like so many scientists before him, Dart seems to have started out with a theory that he wanted to confirm, and to have gathered together the "facts" afterward.

It is not likely that there was any conscious fraud involved. Dart was simply a victim of his own preconceptions. Once he formed the killer-ape hypothesis, he seems to have interpreted the evidence in such a way that his theory was "confirmed" every time new evidence was discovered. He was almost certainly not aware of what he was doing. Scientists rarely realize how their predispositions are leading them astray.

It is somewhat more difficult to explain why, for a time, Dart's ideas were so influential. To be sure, Ardrey's very well-written and interesting popular books were a factor. But there must have been more to it than that. Otherwise, Dart's theories would not have captured the imaginations of people all over the world.

I suspect that the political conditions of the early 1960s might have had something to do with the success of Dart's and Lorenz's ideas. When Ardrey began to popularize the killer-ape hypothesis in 1961, the Cold War still lingered on. The specter of Hiroshima was etched a little more deeply in our minds than it is today. Less than a decade had passed since the first hydrogen bomb had been exploded. For the first time in history, doubts were being expressed about the future of the human race.

Now that another two decades have passed, we have no less reason to feel afraid. But we have become more accustomed to living under the threat of nuclear destruction. If Dart were making his discoveries and constructing his theory today, his ideas might not have the same wide appeal.

Dart's theories about australopithecine violence were questionable, to say the least. But it does not follow that there is no such thing as innate human aggression. The idea that the violent propensities of human beings should be attributed to our ancestor's use of weapons may be erroneous, but it would be absurd to believe that this demonstrates that human violence is not a problem. When doubt was cast on Dart's theories, human beings did not stop making war.

So, we must still consider Lorenz's idea that there exists an innate aggressive drive in human beings. And, even though we now know that animals kill one another too, the idea is not so easy to dispose of. In fact, it may contain a kernel of truth.

However, Lorenz may have overstated his case. Compared to certain other animals, human beings are not especially violent. Although such things as assault and murder happen, they are not typical human behavior. In most cases, interpersonal aggression never escalates beyond the verbal level. We seem to prefer to assault one another verbally, rather than threaten one another physically, as apes and baboons will do.

Our propensity to wage war remains a problem. But it is not

obvious that warfare and individual aggression really have much to do with one another. Even Lorenz now inclines to this view. In an interview that was published in 1974 in *Psychology Today*, he stated that

> If I were to write *On Aggression* again, I would make a much stricter distinction between individual aggressivity within a society and the collective aggressivity of one ethnic group against another. These may very well be two different programs. . . . I may have been wrong in not distinguishing precisely enough between the two factors.

Some anthropologists would go even further and deny that warfare has roots in an aggressive drive at all. "Generals, removed far from the battlefront," says Ashley Montagu, "give orders for annihilation of the 'enemy' with no more aggressiveness or emotion than when they order the gardener to mow the lawn a little closer." The French biologist Jean Rostand comments that "in war, man is much more a sheep than a wolf. He follows, he obeys. War is servility, rather—a certain fanaticism and credulity—but not aggressiveness."

War is not universally practiced by primitive peoples. Eskimos do not make war. Neither do the !Kung* Bushmen of Africa. Primitive peoples who do wage war often do so in such a way that casualities are minimized. The Dugum Dani of the Balien Valley in New Guinea use unfeathered arrows when they do battle; the use of such arrows makes it difficult for them to hit their targets. The Tsembaga, another New Guinea tribe, uses unfeathered arrows also. In Australia, the Murngin fight with spears that have been blunted by the removal of the stone tips, and stage ritualized battles whose object is to settle conflicts without bloodshed. Some Australian tribes aim only for the legs when they throw spears at their enemies.

*The exclamation point in *!Kung* designates a click sound that has no phonetic equivalent in English.

Other primitive tribes throw spears at one another while remaining just out of range. In such "wars," few individuals are injured. Some tribes do seriously try to kill enemies, but stop fighting once a certain limited number of deaths have occurred. There are even peoples who have stopped waging war because the introduction of Western firearms has made warfare too destructive.

When American Indians warred with one another, they frequently confined themselves to raiding one another's camps and stealing horses and women, while avoiding situations where they would have to kill members of other tribes in large numbers. In many tribes, killing and scalping were not considered to be especially creditable acts. The bravest act that a man could perform was to touch or strike an unhurt man and leave him alive. It is true that there were cases where the Indians massacred whites (although, as we all know, the reverse happened more often). But they began to do this only after they discovered that white men were unwilling to fight by "civilized" rules.

Before modern nations engage in warfare, they subject their fighting men—and their civilian populations as well—to intense indoctrination. One of the prime objects of military training is to accustom soldiers to following orders blindly. The soldier and civilian alike are encouraged to view the enemy as less than human. Enemy soldiers are not people, but "krauts," "japs" or "gooks." Atrocity stories are circulated, and every effort is made to convince people that the war is being fought to further some noble cause. Sometimes, of course, the indoctrination effort fails. This is what seems to have happened during the American involvement in Vietnam.

Even when bitter fighting goes on for years, certain rules are observed. The belligerents sometimes refrain from using certain types of weapons, such as gas and disease-carrying microorganisms. Prisoners of war are treated in humane ways—at least, they are supposed to be. Failure to observe the rules of "civilized warfare" is one of the few things that arouses real fury in wartime. In this regard, it is worth noting that many of the protests against the war in Vietnam centered around the use of napalm, a substance that sticks to human skin and develops a temperature of 2,000 degrees centigrade. During World War II and after, reaction against Nazi war crimes (a very revealing term, incidentally) was so great that many of us still find ourselves reacting emotionally when we are confronted with symbols of Nazism today.

Atrocities are practiced in wartime—and in times of peace too—but we do recognize them as atrocities. Human warfare is characterized, not by all-out aggression, but by an insistence that the fighting be conducted in such a way that limits are placed on the amount of physical suffering that it causes. Sometimes we are not very successful when we attempt to do this. But it is an ideal.

Today many of us are all too aware of the fact that humanity is threatening itself with nuclear destruction. However, the cause of this unpleasant situation may not be any innate aggressive drive. It may actually be our propensity for cooperation that makes collective aggression possible.

Our australopithecine ancestors may or may not have been hunters, but in either case, they were social animals. If they did manifest any tendency toward solitary living it soon would have been eliminated by natural selection. Today, a solitary baboon does not long survive; the leopards see to that. The australopithecines must also have been forced to live in groups as a defense against predators.

Humans are still social animals today. They bind themselves in groups and identify with group interests. They are innately so cooperative that they will often place group welfare above individual interests. I tend to think that our propensity for intergroup conflict is more closely related to this than it is to any aggressive drive. In any event, it seems clear that wars arise, not so much out of feelings of anger toward the enemy, but from our too-frequent willingness to die for a cause, for our country, or for our tribe.

Of course, this is only speculation. Our understanding of human nature has not advanced to a point where we can be certain about the causes of our more unpleasant behavioral characteristics. We can't even be sure that war is not a product of cultural evolution, rather than an innate biological tendency. The only thing that is certain is that we have developed a behavior pattern that is becoming more maladaptive with each passing decade.

Biological extinction has been the fate of the great majority of the species of animal life that have evolved on this planet. Some of them became extinct because more efficient competitors invaded the environmental niches that they inhabited. A few died out because they found themselves headed toward evolutionary dead ends. An example of an animinal in this last category is the so-called Irish elk (which was actually a species of deer, not an elk). The Irish elk evolved huge antlers that sometimes measured as much as thirteen feet across. The trend continued because it was related to male competition for females. Only the bucks with large antlers had a chance to mate. Eventually the expected happened. The antlers became so large that they were maladaptive, and the species died out.

It may be that we will suffer a similar fate. At one time our tendency toward intergroup conflict did little harm, at least not from an evolutionary standpoint. Sporadic conflict may have served the biologically useful function of spacing out groups of human beings and preventing the depletion of food resources. In those days, a little fighting every now and then may actually have reduced the frequency of homicidal conflicts between neighbors.

But today this tendency of ours has become as much of a burden as the enormously heavy antlers of the extinct Irish elk. With every passing year, it seems increasingly likely that, sooner or later, we will experience a similar fate.

8

Innate Behavior

Aggression is observed in fish, in reptiles, in birds and in mammals. It exists because it has an adaptive function. If it did not, it would not be seen in so many different species. Aggressive behavior exists because it contributes to the evolutionary fitness of the animals that exhibit it. Otherwise it would rapidly be eliminated by natural selection.

In the previous chapter, mention was made of one of the functions of aggression. It was observed that many animals are territorial. They attack members of their own species who infringe upon territories that they have staked out for themselves. Territoriality is beneficial because it limits population density. Environmental resources—especially food—are not depleted.

A good analogy is provided by the activities of professional organizations like the American Medical Association. AMA restricts the number of physicians by limiting the number of spaces available in medical schools. By ensuring that only so many "territories" shall be available, and by preventing the number of doctors from increasing too rapidly, AMA protects their social prestige and income. If there were too many doctors, and if they had to engage in fierce competition for patients, they would suffer a fate analogous to that of a species of animal that bred too rapidly and outran its food supplies.

It is true that natural selection knows nothing of "the good of the species"; it acts only on individuals. However, in this case, individual and species interests coincide. An individual animal is better off if it maintains a territory and does not allow other animals access to resources on that territory. The species is better off if population density is not too high.

Some animals—seals, for example—are territorial only during the breeding season. Seals compete with one another, not for physical territories that provide them with food, but for breeding sites. The males fight with one another to obtain portions of secluded beaches. Those that fail to do so are unable to mate with the females and father offspring.

The biological fitness of animals that are unable to mate is zero. However large and well fed an individual may be, it cannot be said to be fit if it does not pass its genes along to the next generation. Thus, breeding territories are as important a resource as territories that afford an animal access to food.

Some animals—such as cheetahs, mountain goats, deer, wallabies, rhesus monkeys and baboons, to mention just a few species—are not territorial at all. Yet these animals also exhibit aggressive behavior. It is obvious that territoriality is only one of the uses of aggression.

Some of these species fight for mates. Deer and mountain goats, for example, engage in combats with one another. Deer wrestle with their antlers, and mountain goats run at one another and butt each other on the head. The vanquished animals are never killed and are only rarely injured. However, since the females will mate only with the winning animals, the losers are denied an opportunity to reproduce. The biological benefits of this are obvious; only the strongest animals are able to pass along their genes to the next generation.

Animals will fight for territorities and for the opportunity to mate. But this does not exhaust the list of the uses of aggression. Aggression has yet another important biological function.

Before I explain what this function is, it might be best to say a few words about what it is not. Aggression—at least in the sense that we are talking about—has nothing to do with preying behavior. When carnivorous animals hunt, they are no more engaging in aggressive behavior than humans are when they go fishing or slice up the Thanksgiving turkey. An animal that is stalking its prey does not exhibit anger. It does not engage in threatening behavior that is designed to intimidate its opponent. A leopard that is chasing its dinner exhibits behavior that has nothing in common with that of an animal that is snarling with rage.

Of course, it would be possible to make a case for calling predatory behavior "aggression" if we were so inclined. Words, as Lewis Carroll's Humpty Dumpty pointed out, can be made to mean whatever we define them to mean. However, preying and intraspecific fighting are so unlike each other that most zoologists prefer to use the term *aggression* to designate only the latter.

If this use of *aggression* has nothing to do with preying, then what is it? The answer is far from obvious. In fact, the first time that one hears it stated, it sounds somewhat paradoxical: animals fight in order to limit the amount of fighting that they must engage in.

They do this by setting up dominance hierarchies. Animals that live in groups will fight to establish dominance over one another. But once dominance relations are established, there is little fighting. The alpha male in a group of baboons does not have to constantly engage in personal combat to maintain his status. Once he has established himself at the top of the hierarchy, he generally needs to do nothing more than fix his subordinates with threatening stares whenever he needs to remind them of their lower status. In other words, baboons fight so that they won't have to be constantly fighting.

Naturally dominant males are sometimes overthrown. However, this happens less often than one might think. High-ranking baboons frequently ally themselves with one another and act in unison to repel any threats to the existing hierarchy. In one baboon troop observed by the American anthropologist Irven DeVore, the central hierarchy consisted of three males named Dano, Pua and Kovu. But the strongest individual was a fourth male, named Kula. DeVore found that Kula was nevertheless obliged to move aside whenever he encountered Dano, Pua and Kovu together.

A somewhat similar situation has been observed in a troop of Japanese macaques. This troop was headed by a single male, named Arrowhead. The twenty-five-year-old Arrowhead was elderly by macaque standards, was slightly built, had lost three of his four canine teeth, and was blind in one eye. And yet he was able to maintain his dominant position. He was kept in power by a group of allies who repelled any challenges.

Dominant males (and dominant females too; females often have their own dominance hierarchies) enjoy a number of advantages. They have access to any choice morsel of food that a troop runs across. Dominant animals will not hesitate to take food from subordinates. When food is relatively scarce, during times of drought for example, the latter may not get enough to eat. Dominant males also have certain breeding privileges. When a female baboon first comes into estrus, the subordinate males will be allowed to copulate with her. However, she will be monopolized by a dominant male during that part of the estrus period during which she is most likely to conceive.

If dominant males have privileges, they also have responsibilities. High-ranking baboons maintain order in their troops by quickly breaking up fights that break out among subordinates. They also defend their troops against predators. When a leopard approaches a group of baboons, an alpha male will rush forward and threaten the cat with its bared canines. Although the leopard is perfectly capable of killing the baboon it is facing, it will normally retreat rather than risk injury.

Baboon alpha males are not the only animals that look out for the welfare of their subordinates. Some very revealing observations were made in the middle 1960s when some primatologists released a colony of woolly monkeys into an area that was unfamiliar to them. They found that the alpha male would not permit other members of the colony to climb into trees until he had inspected routes for climbing and had broken off the dead branches. The subordinate animals were permitted access to the trees only after the dominant male had spent two days engaging in careful exploration.

Whenever we hear such stories, we cannot help but find them intriguing. The reason is obvious enough: they remind us of human behavior. Groups of human beings have their dominant animals too. Every organization, every association or club, every office where people work together has its hierarchy. Hierarchies exist in social groups, in political bodies, in military organizations, even in children's play groups. In families, parents dominate the children, and the oldest child will frequently dominate its siblings.

There is ample evidence that we humans are at least as concerned with dominance rank and status as are apes and monkeys. The only way in which our behavior differs from theirs is that our larger and more complex brains make it possible for us to be more subtle about it. We do not threaten our subordinates by baring our teeth, and we practically never have to fight in order to maintain privileged positions.

Human beings have invented numerous methods for proclaiming their positions in the pecking order. We do it with the clothes we wear, with the cars we drive, with the houses we live in. An address in a fashionable neighborhood can be an indication of a high position in the dominance hierarchy. Status is reflected even in the language we use. The ability to write a literate business letter, for example, is a sign of high status, while the use of certain "illiterate" colloquialisms is an indication of a low one.

A dominant animal—whether it is a human being or a baboon—must not give the appearance of being too insecure about its status. If a baboon is always nervously threatening its subordinates as though it was anxious about its ability to maintain order, the chances are that it will soon be overthrown. An alpha male who really feels secure may do a little threatening just to spread terror, but when he wants to he can intimidate his subordinates simply by fixing them with an arrogant stare.

Humans who wish to be accepted as dominant animals must behave similarly. In every age, the *nouveau riche* fail to gain acceptance because it is obvious that they are trying too hard. Wearing a fur coat to the supermarket is not a sign of high status. Wearing a pair of jeans with an air of unconcern can be.

It is obvious that cultural factors affect the forms that dominance displays take. An American businessman, a member of the Saudi Arabian royal family and a New Guinea chief will display high status in different ways. However, striving for dominance and exhibiting it once it is achieved seem to be innate human behavioral traits. Every human society has its hierarchies. In every one, high-ranking members of the hierarchy want to make it known that they occupy important positions.

A high-ranking animal is always the center of attention. Baboons constantly pay attention to what the alpha male is doing, and they follow him when he decides to go off in any direction. Humans act the same way. They listen to what their high-ranking males and females have to say, and sometimes even imitate their modes of dress or behavior. When children play, they will follow the suggestions of the most dominant individuals and will ignore the ideas of those who are subordinate in their hierarchy.

Since human beings are social animals, dominance relations are necessary. A human society could no more function without a hierarchy than a baboon society could. Without hierarchies, there could be little cooperation between individuals; we would not know whom to follow. A group of friends who go bowling once a week needs a hierarchy; so does an army or a corporation. Even the Arapesh, who discouraged competition within their society, found that they had to persuade certain individuals to act as big men. It was the only way to see that things got done. Whether one is fighting a war, gathering pigs for a feast, making decisions about marketing a product, or deciding what bowling alley to go to, it is necessary to have one or more dominant individuals who will get everything organized.

High-ranking baboons have access to the best food. They have certain mating privileges as well. In the human species, the benefits that accrue to the individuals at the top are similar. Dominant males have high incomes. So do dominant women. Those of the women tend to be a little lower, because males still dominate females in our society. One symptom of this is the fact that women are most attracted to dominant men, while men are more interested in women's physical attractiveness. Of course, there is more to it than that; human behavior is always complex. In particular, a woman's attractiveness is not unrelated to her position in the hierarchy. A woman of high socioeconomic status will be likely to have the financial resources and knowledge that will allow her to select clothing that enhances her appearance, and to be skilled in personal adornment. After all, one of the signs of a high position in the hierarchy is the possession of "good taste."

When we look at the behavior of animals, especially that of our closest relatives, the other primates, we often see behavioral traits that have counterparts in human beings. When we observe dominance hierarchies in troops of baboons, we are reminded of the human quest for status. When we read of chimpanzees sharing meat (chimps do hunt; they prey on monkeys, young baboons, bush pigs and bushbucks, among other animals), we think of the human custom of inviting friends over for dinner. In such cases, we take particular care to select a choice piece of meat, such as a roast, for the occasion.

There may even be similarities in the use of "obscenity." In some species of monkey, males will threaten other males with genital displays. Squirrel monkeys display the erect penis to assert dominance over their inferiors. Common marmosets defend their territorities by raising their tails and presenting their rears to their opponents. An erection then takes place. Admittedly, humans do not engage in behavior that is quite so outrageous (outrageous for humans, that is; it is quite natural in monkeys). However, some Papuan tribes wear ornaments that emphasize the penis, and the codpiece was once part of the apparel of high-ranking males in Western society.

The similarities between penile displays in monkeys and certain kinds of behavior in humans could very well be accidental. After all, there is no evidence that our hominid ancestors used such displays, and they are not observed in apes. However, if the similarity is not accidental, it could explain why references to copulation so often become verbal insults and threats.

And what about the obscene gesture that is known in the United States as "giving the finger"? This threatening gesture, which is used by both men and women, is made by extending the second finger and bunching the thumb and fingers on either side. Its meaning is obvious. The extended finger represents the erect penis, while the thumb and first, third and fourth fingers correspond to the scrotum. This same gesture, by the way, was used in Roman times, when it was known as the *digitus impudicus*.

It is possible to see many similarities between human and primate behavior. But, of course, there are also many ways in which humans and primates are different. Human language and culture have led to the development of numerous forms of behavior that have no counterparts in the animal world. There is also a difference that has nothing to do with language or culture or technology. This is the uniquely human preoccupation with sex.

It has sometimes been observed that man may be the sexiest animal in existence, and much has been made of the size of the human penis. An erect human penis, for example, is much larger than that of a gorilla, even though the gorilla is the larger animal.

There are some reasons for viewing such comments with skepticism. For example, the large size of the human penis may be nothing more than an adaptation to the changes in the size and positioning of the vagina that came about when our ancestors adopted an upright stance. And it is not entirely clear what is meant when it is said that human beings are the "sexiest." For example, are human women "sexier" than the female chimpanzee who may copulate with forty different males in a single day? Are human males "sexier" than the lion who copulates a hundred times in a twenty-four-hour period?

However, it is clear that humans are preoccupied with sex in a way that no other animals are. In all the other primate species, females are sexually receptive only during a relatively short period of estrus. With a few rare exceptions—orangutans, for example, have been observed to commit rape—primate males show no interest in sex when no females that are in estrus are about.

Humans, on the other hand, exhibit an almost continuous interest in sex. Humans create pornography, describe sexual encounters in literature, and use sexual symbols in art. Men fantasize sexually about strange women they see on the street. Sex is used in advertising. There are sexual jokes. Sex even plays a prominent role in religious scriptures. Hindu writings describe the loves of Krishna in great detail. An erotic poem, "The Song of Solomon," is part of the Old Testament. Christian mystics have used sexual symbolism to describe their highest mystical experiences.

It is obvious that this preoccupation must have something to do with the fact that human females are the only primates that are sexually receptive at all times. Other female animals go into estrus for brief periods, and once they conceive, all interest in sex ceases. Years may pass before they begin to solicit males again. Female chimpanzees will not experience estrus for two or more years after the birth of an offspring. This has the effect of spacing out chimpanzee births and preventing a situation in which a female has more infants than she can handle.

Human females, on the other hand, remain interested in sex during the infertile periods in their menstrual cycles, while they are pregnant, and while they are nursing their infants. Almost certainly it is this that has made us so sexually preoccupied. With the exception of males who have become impotent for one reason or another, every adult human being is capable of engaging in sex at any time. Under such circumstances, it would be very difficult not to think about the activity.

But why is the human female continuously receptive? It is obvious that female receptivity must have provided the human species with some unique evolutionary advantage. Otherwise, this odd behavior pattern would not exist. Like the other living primates, we are descended from species in which the females went into estrus. Why should natural selection have caused estrus to disappear only in humans?

This is not an easy question to answer. After all, we do not know exactly when estrus disappeared in our ancestors. Neither estrus nor continuous sexual receptivity leave fossils. We do not know what hominid society was like when the change took place. Therefore, if we are to come up with an answer, it is necessary to engage in a certain amount of guesswork.

One theory that has gained a certain amount of prominence is widely—but not universally—accepted today. It was first proposed during the 1960s and was expounded in a number of popular books, most notably in Desmond Morris's *The Naked Ape*. This theory relates continuous sexual receptivity to the human tendency toward what is known as *pair bonding*.

A pair bond is nothing more than an emotional bond between a man and a woman who have a sexual relationship. Pair bonds exist in every human culture. In other words, long-term sexual relationships are a universal human custom. The free promiscuity that nineteenth-century anthropologists thought was a stage in cultural evolution never existed—at least, not since the time that we evolved into *Homo sapiens*.

Pair bonding is not the same thing as monogamy, however. Human beings are perfectly capable of forming bonds with more than one individual. Something like 80 percent of the human societies that have been studied are polygamous, and there are sexual relationships outside of marriage in monogamous and polygamous societies alike. True, strict monogamy is sometimes considered to be an ideal in Western culture. But this is due to the influence of Christianity.

Pair bonds do not always persist for life. Divorce is common in many societies, and it is often economic factors that cause couples to stay together in primitive and technologically advanced societies alike. In the United States many people remain married because they feel responsibilities to their children, or because divorce, in some segments of our society, still meets with social disapproval. There are many reasons why a couple may remain together when strong emotional bonds no longer exist.

Finally, pair bonds do not necessarily have anything to do with the intense feeling we call "romantic" love. In some societies, marriages are arranged, and emotional attachments develop after marriage, not before. Even in our own culture, people frequently marry for reasons that have little to do with the romantic variety of love. They may be more interested in companionship, for example.

And of course, pair bonds can develop between a man and a woman who have not yet had sex. They can come into existence between two men or two women as well.

Pair bonds are of such varying duration, and are formed in so many different ways, that it is difficult to make generalizations. However, one thing seems clear. A propensity for forming pair bonds is innate. If it were not, it would not be found throughout the species. In other words, humans have their own characteristic mating system. In most cases, they are not freely promiscuous like chimpanzees. They rarely exhibit the exclusive, lifelong bonds that are common in so many species of birds. Instead, they form bonds that may or may not be sexually exclusive and which are of varying duration.

It seems plausible to assume that the absence of estrus and the existence of human pair bonding are related. After all, human bonds are strengthened by frequent copulation. In many cases, bonded individuals lose interest in other sexual partners, at least for a time. Human pair bonds are sexual in nature.

If we accept the idea that continuous sexual receptivity exists in human females because it makes strong pair bonds possible, we really haven't explained very much. In fact, we have simply replaced one question with another, going from "Why has estrus disappeared in human beings?" to "Why do human beings form sexual bonds?"

According to the theory I have referred to, pair bonds provide human beings with a distinct evolutionary benefit, one that is related to the relative helplessness of human infants. According to this theory, pair bonds induce males to aid in the care of their offspring. Pair bonds evolved because they cause males to provide protection to their mates and infants, and to share food that was obtained by hunting. In addition, the existence of pair bonds supposedly reduced the amount of fighting over females. This would have made it easier for the males to engage in the kind of cooperative behavior that was required if hunting expeditions were to be successful.

When the theory was first proposed, quite a bit of emphasis was placed on the hunting behavior of the early hominids. At the time, it was believed that meat constituted a large part of the diet of these hominids. Consequently, it was thought that hunting behavior produced evolutionary pressures that led, in turn, to the rapid evolution of the human brain. Hunting, it was suggested, demanded greater intelligence than the gathering of food plants.

In the last decade or so, serious flaws have appeared in this argument. It has become apparent that hunting may not have been so important to the survival of our hominid ancestors as was thought. Anthropologists began to realize that, in the hunting-gathering societies that still exist, it is the women, not the men, who provide most of the food. The vegetable foods that the women gather account for something like 80 percent of the total caloric intake. Furthermore, hunting is not the only activity that adds animal protein to the diet. The women gatherers collect not only plants, but also such foods as frogs and birds' eggs.

Certainly hunting would have enhanced our ancestors' chances for survival. After all, there are periods during which staple plant foods become relatively scarce. A more or less continuous supply of meat would certainly have made life a little easier. However, hunting seems not to have been absolutely necessary for survival. Therefore the theory has been somewhat modified. Pair bonds, it is now asserted, induced both males and females to share food. Meanwhile, the idea that males were induced to provide protection and parental care is retained.

It must be admitted that it sounds very plausible. At some point in our history, all females certainly went through estrus, just as the females of other primate species do. Furthermore, there would certainly have been some variation; some prehuman females would have been receptive for longer periods than others. If the latter were more likely to obtain male protection, and to be given meat, they would produce more offspring. Natural selection would cause the genes for longer female sexual receptivity to spread throughout the population. Sooner or later estrus would disappear.

As I said, this is a plausible theory. But it is not necessarily true. Plausibility and proof are two different things. One of the difficulties is that there is no way of knowing whether or not estrus in hominid females disappeared in the way that the theory requires.

It is known that our ancestors were engaging in hunting at least two million years ago. If there is a connection between hunting and the disappearance of estrus, continuous female receptivity must have evolved at about the same time. It is easy enough to determine that the early members of the genus *Homo* hunted. The tools and fossil bones of butchered animals that have been discovered with hominid remains are sufficient to establish that. But there is no way of knowing what their mating system was like.

Science has known numerous plausible theories that were eventually discredited. However reasonable they seemed, they had to be given up if they were contradicted by the evidence. Unfortunately, the pair-bond theory seems to be susceptible neither of proof nor of disproof. As long as we cannot travel back in time and observe what early hominid society was really like, there is no way of knowing whether its assumptions bear any relation to reality.

Some scientists have become skeptical about the theory in recent years. Primatologists have made a number of observations of primate behavior that have caused them to doubt its validity.

Although our fellow primates are our closest relatives, and although humans have always been interested in them, relatively little has been known about the behavior of most primate species until fairly recently. The reason is that most species live in relatively inaccessible habitats. It is true that some primates—baboons, for example—are not hard to study; one needs only to drive out on the African savannah and watch them. However, it is quite another matter to tramp through a remote jungle to study a rather inconspicuous kind of monkey.

Of course, one can always study the behavior of primates in zoos. Unfortunately, observations made under such conditions are of doubtful validity. Animals often exhibit abnormal behavior when they are in captivity. If one wants to know how they behave under more natural conditions, one has to go out and study them in their natural habitats.

In the last few years quite a bit of significant data has been collected, and some of it has a direct bearing upon the pair-bond theory. In particular, it has become apparent that male members of all primate species give some kind of care or protection to infants. Male Japanese macaques carry year-old infants about when their mothers are giving birth to new babies. Male gorillas will defend juveniles against predators and will look after orphaned infants. And although most male primates do not share food with either infants or females, such sharing has been reported among some monogamous primate species.

Male solicitude sometimes expresses itself in quite dramatic ways. For example, there is the case of a patas monkey who pursued a jackal and snatched away an infant that the predator was carrying off in its mouth. And then there is the langur male that rescued an infant from a raptor that had swooped down and had nearly caught a baby in its talons. There has even been an observation of a baboon assisting at a birth. It seems that a female hamadryas baboon had gone into labor while on a steep cliff near her sleeping site. As the baby emerged, the mother's rump was protruding over a precipice. For a moment, the infant was dangling by its umbilical cord over the side of the cliff. It would almost certainly have fallen to its death if a male baboon had not rushed over. He reached out and caught the infant and handed it to its mother.

It is the parental behavior of primate males who are not rewarded by continuous female sexual receptivity that makes one wonder whether the human pair-bond theory can be correct. It is likely that human females would have been able to count on a certain amount of male protection for themselves and their offspring even before natural selection did away with estrus.

Furthermore, approximately 18 percent of the various species of primates are monogamous. They form pair bonds that are maintained even though the females remain sexually unreceptive for long periods of time. In many of these species—which include some lemurs, tarsiers, and Old and New World monkeys—the males will either assist in the carrying and feeding of infants, or defer to mates and offspring over food.

Such observations have the effect of making the human pair-bond theory seem somewhat contrived. If natural selection can shape male behavior in other primate species in such ways, there is little reason to conclude that loss of estrus was really necessary to produce food sharing and parental behavior in human beings. When pair bonds and male solicitude are beneficial to survival in a primate species, natural selection is perfectly capable of bringing them about by a less roundabout route.

If continuous female receptivity does not exist to elicit male participation in the nurture of offspring, then what purpose does it serve?

I do not claim to be able to answer this question. I am no more able to travel back in time to see how they arose than the scientists who devised the pair-bond theory. However, I think that I can show that other plausible theories are possible. I propose to do this by constructing just such a hypothesis. I do not claim that it is the correct one, but I believe that it has just as much chance of being true as the pair-bond theory.

First, let us ask whether there really is a connection between the loss of estrus and the existence of human pair bonds. Once we begin to entertain a little skepticism about the pair-bond theory, it is easy to see that it is by no means obvious that any such connection exists. Pair bonds and estrus exist simultaneously in many primate species. In those species, pair bonds are related to parental behavior. But continuous female receptivity is not, for the simple reason that continuous receptivity does not exist.

Then why does continuous sexual receptivity exist in human females? One could presumably answer that question by saying, "That's just the way it happens to be." But that would hardly constitute a scientific explanation. In fact, it isn't an explanation at all. The loss of estrus in the human species is unique; there must have been some reason for it.

Perhaps the best way to begin our speculation would be by examining the phenomenon of infanticide in lions. A pride of lions typically consists of one or two adult males, many females, and their juvenile offspring. There also are solitary male lions that are not members of prides. If these solitary males are to have any chance to reproduce, they must gain possession of a pride of females by driving off or killing the resident males.

When new males take over a pride, they kill the females' cubs. If any females are pregnant, the new males will also kill the cubs as they are born. It is obviously to their biological advantage to do so. Their fitness will be enhanced if the pride's food resources are invested only in their own offspring.

Primatologists have recently become aware that infanticide is also common in many species of primates. It is only during the last few years that such cases of infanticide have been well documented. Scientists had heard stories of such behavior for years, but until they actually observed infanticide themselves, they tended to discount such tales as mere folklore. It seems that they paid less attention to eyewitness accounts that were given by natives than they did to preconceptions about "cooperative" primates that always acted for the common good. In fact, when the existence of primate infanticide was first documented, some scientists argued that it had to be a pathological aberration brought about by crowding or by human disturbance.

There is still a certain amount of controversy on the subject. In particular, no one is really quite sure how common infanticide is. However, it does seem to occur among the great apes, New World monkeys and cercopithecines, one of the families of Old World monkeys. It may occur in other primate species as well.

Chimpanzees are one of the species in which infanticide has been observed. Chimps do not attack or kill infants that belong to their own groups. But they will kill the offspring of strange females. The best-documented case is one that was observed at the Gombe

Streal Reserve in Tanzania in 1971. Primatologist David Bygott had been following a group of five chimpanzees. When the animals suddenly encountered a strange female from another community, they immediately attacked the female and her infant. A few minutes later, one of the chimps was holding the infant's legs and intermittently beating its head against a branch. After a few more minutes had passed, he began to eat the flesh from the infant's thighs.

Since Bygott made this observation, Jane van Lawick-Goodall has reported seven additional cases of infanticide at Gombe. In three of these cases, the "murders" were committed by adult females. In the other four, the infants were killed by males. In all of the cases involving males, the victims were the offspring of strange females. Male chimpanzees apparently do not kill infants belonging to females that are members of their own community.

Since chimpanzees mate promiscuously, another way of saying the same thing would be to state that male chimpanzees do not kill the offspring of females with whom they have mated. In fact, they are quite tolerant toward and solicitous of infants and juveniles of their own group.

At first glance, infanticide hardly seems to be a very adaptive kind of behavior. Can such a thing really have evolved by natural selection?

The answer to this question is Yes, of course. If we recall that natural selection acts on individuals, not on species, it is easy to see how it might have come about. A male who kills the offspring of other males is enhancing his fitness. The reduced competition makes it more likely that his own offspring will survive. Furthermore, in some species, such as chimpanzees, females will not go into estrus while they are still nursing. A female whose infant is killed will come into estrus again that much sooner. If she mates with the animal who killed her infant, he will have more offspring than he would have had if he had not committed infanticide. This is a common motivation for infanticide in monkeys. When a new male takes over a group, he will frequently attempt to kill the offspring of the old alpha male.

The existence of infanticide in primates also provides an explanation for certain curious kinds of behavior. For example, it has long been known that savannah baboons will often pick up infants when they are engaged in confrontations with other adult males, especially with males that have recently immigrated into the troop. Until fairly recently, primatologists explained this behavior by appealing to a presumed baboon inhibition against harming baboon babies. According to this theory, the baboons picked up infants to use them as shields; they wanted to inhibit violence against themselves.

This theory was exploded in 1980, when two scientists, Curt Busse and W. J. Hamilton, engaged in an experiment that involved sedating baboons to obtain blood and other samples. Busse and Hamilton discovered that, when a mother was drugged, immigrant males would take advantage of the opportunity to kill her infant. It suddenly became obvious that baboon males did not pick up infants to shield themselves. They did this to prevent the infants from being killed.

At the moment, it is not possible to say whether infanticide is practiced by the males of most or all of the nonhuman primate species, or only in some of them. The killing of infants is frequently difficult to observe; in many cases the evidence is only circumstantial. If a primatologist notices that an infant that has previously been attacked by an adult male has suddenly disappeared, he may suspect that the male has succeeded in killing it. But, of course, he cannot be certain; the infant might have been carried off by a predator.

One might ask why females do not prevent the males from killing their infants. Well, in many cases they try to. Sometimes they are successful. However, it would not enhance the fitness of the female to fight too fiercely. If an infant is killed, she can always have more. But if she is killed or severely injured, then her child-bearing career will be over. In many cases, the behavior that most enhances the female's biological fitness will be to mate with the same male that has killed her infant.

Tales of animal infanticide sound grisly to human beings. One of the reasons that they do is that male infanticide is relatively rare in our own species. When infanticide is committed, most often the infant is killed by, or with the consent of, the female as a form of birth control. The killing of babies by strange males—in war, for example—is much less common.

Human males don't often commit infanticide. Human females are continuously sexually receptive. Neither one of these traits seems to be typical of primates. Could it be that they are somehow connected?

Suppose that, at one time, hominid males did kill the offspring of strange females. Suppose also that, like lions, they killed the offspring of females with whom they had never copulated when the latter gave birth. There is nothing unreasonable about either assumption. As we have seen, the infants that other animals kill are those that could not possibly be their own offspring.

Now suppose that some hominid females became receptive for short periods after they were already pregnant. This is not an unreasonable assumption either. As a matter of fact, short periods of pseudo estrus have been observed in pregnant chimpanzees.

Mothers who exhibited sexual receptivity during pregnancy would have a biological advantage over those who did not. A male who had copulated with a pregnant female would be less likely to kill her child when it was born, because he could not be sure that it was not his.

For that matter, continuous sexual receptivity might also have been a factor in inhibiting males from killing infants that had already been born. In this connection, it is interesting to note that human beings may be the only animal species in which males will care for infants that cannot possibly be their own.

In other words, continuous sexual receptivity may have evolved, not to induce males to share food and engage in parental behavior, but to prevent them from killing infants that they had not fathered. Such a situation would not do much to enhance the males' fitness, but it would certainly be beneficial to that of the females, who had just as much interest in passing their genes along to the next generation.

This theory is not as pleasant to contemplate as the pair-bond theory. However, I do not think that it is implausible. After all, male mortality rates among our hominid ancestors were probably rather high. If the females who had lost their mates had some way to prevent new mates from killing their children, it would certainly have provided an evolutionary advantage. In fact, prevention of infanticide might have been necessary before human beings could have evolved. The period of infant dependency is much longer in the human species than it is in any other. This makes infants relatively more vulnerable than they are in ape species, or were among our smaller-brained ancestors.

This long period of dependency would explain why estrus disappeared in humans, but not in the other primates. Since a monkey or an ape grows up much more quickly, its period of vulnerability is shorter. Thus, prevention of infanticide is not quite so vital.

Perhaps this would be a good point at which to bring the speculation to a halt. It is no easier to tell whether our hominid ancestors were killing infants two or three or five million years ago than it is to tell when estrus disappeared in females. I can think of no way in which the theory I have outlined could be substantiated. My only intention was to show that it was plausible, and that we ought to remain skeptical about the pair-bond theory, which might be somewhat less plausible.

If the pair-bond theory can't be proved either, why has it been so widely accepted? Anyone who has followed the arguments that I have been making throughout this book should have little difficulty in suggesting a possible answer to that question. The pair-bond theory gained acceptance because it fit scientists' preconceptions.

Romantic love, after all, is given a great deal of emphasis in Western culture. We constantly encounter love stories in television, in movies, in popular music, in literature, and even in comic strips. What, then, could seem more natural than a theory that emphasized the role of love in human evolution? Once we have made that observation, is it really necessary to add anything more?

Certainly this theory is more appealing than the alternative one that I have outlined. But the fact that a theory is appealing does not make it true. The nineteenth century found social Darwinism to be appealing. As a result, Spencer's ideas were very widely accepted. But this did not imply that they were correct. The idea that human behavior is shaped wholly by environmental factors is one that has always appealed to people with liberal political orientations. But it doesn't follow that this theory is right either.

The world is not always what we would like it to be. Sometimes we react to this fact by pretending that it is something that it is not. In most cases, there is nothing wrong with this. Illusions can be very comforting, and in many cases it does no harm to believe in them.

However, if we wish to approach a subject scientifically, we must attempt to throw our illusions aside. Unpleasant truths should be accepted as readily as those that are emotionally neutral or appealing. Naturally, scientists do not always accomplish this. Since they are as human as everyone else, they cannot help being influenced.

Every human culture maintains certain cherished beliefs that may or may not have anything to do with reality. When scientists do not carefully scrutinize these beliefs, they can be as much influenced by them as anyone else. Sometimes this really makes very little difference. I doubt that it matters very much whether a chemist or a mathematician happens to believe in the importance of romantic love, or whether he is a democrat, a Marxist, or even a monarchist.

However, the study of human nature is an entirely different matter. It is probably unrealistic to expect scientists to proceed in this area without being influenced by preconceptions of numerous different kinds. But at least we can examine their theories and try to see what kinds of preconceptions they might have had.

9

Sociobiology

Although speculations about human origins and studies of animal behavior can be fascinating, there are limits to what they can tell us about human nature. If we place too great a reliance on them, and forget that they are speculations, not facts, we may very easily find ourselves drawing conclusions that are not very accurate.

For example, we know that *Homo erectus* was a hunter, and that he had a brain that was about two thirds as large as that of modern man. We know that he used stone tools, and we even have a pretty good idea what the various tools were used for. But there is not much else that we can be certain of. We do not know the mating customs of *Homo erectus*. We do not know how aggressive he was. We do not know many details of the society in which he lived. We can't even be completely sure that he possessed the use of language.

We can make guesses about the life that *Homo erectus* lived, and we can invent arguments that make these guesses appear plausible. But we cannot travel back into time to confirm their validity. And since we cannot, there are limits to the usefulness of speculation about his behaviorial traits, and about what relevance these might have for modern man.

Comparisons between animals and human beings can be even more misleading. As Margaret Mead pointed out, it is always possible to find at least one species of animal that behaves in just the right way to substantiate virtually any point that one could want to make about human behavior.

In the previous chapter, I compared penile displays in monkeys to obscene insults and gestures that are used by human beings. But is there really any connection between the two? No one knows. The similarity that we perceive could very well be accidental. Even if it is real, it is not obvious what conclusions should be drawn. And what is one to make of the fact that in the human species these gestures and insults are used by both men and women?

Dominance hierarchies are observed in numerous animal species. They are also a characteristic of groups of human beings. Surely this fact has some significance? After all, competition for dominance is something that is found throughout the animal world. It is not something that is seen only in a few obscure species of monkeys.

In this case, I think that we have to admit that there is a real connection. However, it is doubtful that this fact gives us very many new insights into human nature. After all, human beings realized that they were competing with one another thousands of years before anyone noticed pecking orders in hens or in chimpanzees. Furthermore, dominance in animals can be related to social structure, social behavior and mating systems in numerous different ways. For example, when a troop of baboons travels, the dominant males and the females can be found in the center of the group, while the subdominant males walk on the periphery. Does this have any relationship to any of the things that humans do? If so, what?

If we are to have any hope of understanding animal behavior—or any other kind of natural phenomena, for that matter—it is necessary to construct some kind of theoretical framework that will bring everything together. The collection of data is an important part of scientific activity. But it is only a part. Without some theory* to guide us, we can look at animal and human behavior

*By "theory" I mean an explanation that is supported by a certain amount of hard data; when I say "hypothesis" or "speculation," I am referring to something more tentative.

until we are blue in the face, and we will have nothing but a large number of disconnected observations.

Noting similarities between the behavior of humans and the behavior of animals often tells us very little, either about ourselves or about the species we observe. On the other hand, if we can develop a theory that explains why animals act the way they do, we will have reason to hope that the theory can be used to enlighten us about ourselves also. The theory may even be used to explain aspects of human behavior that have no counterparts in animals.

In order to get an idea as to how such a theory might be developed, we will examine yet another kind of animal behavior: rape in mallard ducks. Before I go on, it is necessary to caution that I am not attempting to draw any parallels between this behavior pattern and the crime of rape as it is committed by human beings. The similarity, after all, may be only superficial. For all we know, duck rape has nothing to do with human behavior of any kind. If we want to be truly scientific, we must attempt to understand it in its own terms. If we are successful in this endeavor, we may hit upon a theory that has a much broader range of application.

The males of many species of ducks are rapists. Mallards are especially notorious in this respect. Normally, before mallards mate, they engage in prolonged courting behavior. Copulation of a mated pair follows an elaborate series of behaviors and responses. But sometimes a male will chase after a female that he has not been courting. If he catches her, he will place his beak on the back of her head or neck and force copulation.

The reasons for this behavior are not very difficult to understand. If there is an imbalance in the numbers of males and females who inhabit a pond, rape may be the only way that an unmated male can reproduce. If a mallard drake commits rape, he must be doing it to enhance his fitness (otherwise the behavior would not have evolved). Recall that an animal's fitness is measured not by the length of its life, but by the number of offspring it has.

Rape is not common in the animal world. The males of most

species attempt to maximize their fitness in other ways. Nevertheless the males of some species do rape. When they do, it is not hard to figure out why.

It is the behavior of the raped female's mate that seems, at first, to be puzzling. If he can come to the aid of the female in time, he will attempt to chase her attacker off. There is nothing surprising about that. What is surprising is his behavior when he cannot prevent the rape from taking place. Under such circumstances, he will immediately rape the female himself. He will force copulation without first going through his habitual courtship rituals.

But this behavior begins to seem less puzzling when we consider that the mated male too wants to maximize his fitness. In particular, he wants to lessen the probability that his mate will be fertilized by the strange drake. There is only one way that he can accomplish this. If he inseminates her himself, without first going through the time-consuming precopulation ritual, his sperm will have a chance to compete with those of the rapist.

Of course, the male does not consciously reason all this out. When I say that he "wants" to introduce his own sperm as quickly as possible, I am speaking metaphorically. What he is really doing is exhibiting a behavior pattern that has been shaped by natural selection.

Suppose that, at one time, some mallards had genes which caused them to act in such a way, while others did not. Drakes who raped their mates under such circumstances would have more offspring than those who did not. In time, their genes would spread throughout the population, and the behavior pattern would become universal.

It seems that we have succeeded in explaining rape in ducks by making the simple assumption that any behavior pattern that enhances an animal's fitness will be established by natural selection. Now it so happens that theorizing of this sort has a name. It is called *sociobiology*.

If sociobiology were any normal kind of subject, I would give a definition of it at this point. But it is not so easy to do this. Sociobiology has evoked quite a bit of controversy within the scientific world. Not only do scientists argue about its merits, they can't even agree whether sociobiology is a theory or a discipline!

In 1975, Harvard entomologist Edward O. Wilson published a book called *Sociobiology: The New Synthesis*. Today, this massive work is still sociobiology's major text. Now, Wilson does give a definition. He says that sociobiology is "the systematic study of the biological basis of all social behavior." Personally, I find this a little vague. I prefer the version given by University of Washington zoologist David Barash. According to Barash, it is possible to state the "Central Theorem of Sociobiology" as follows:

> When any behavior under study reflects some components of genotype, animals should behave so as to maximize their inclusive fitness. In most cases this is achieved by maximizing the production of successful offspring.

In other words, whenever behavior has any genetic components, one can explain it by looking for ways in which that behavior might have evolved by natural selection. Since sociobiologists tend to believe that most or all animal behavior is influenced by genetic factors, they seek evolutionary explanations for practically everything.

One can call sociobiology a *theory* if one wants, or one can use the term *scientific discipline*. It probably makes little difference which term one uses. In either case, sociobiology can be described as a scientific activity that revolves around the idea that genetic explanations of behavior are possible. One example of such an explanation would be the one I gave for the actions of the mate of the female mallard.

I have deliberately put off defining Barash's term "inclusive fitness." This term is related to what Wilson calls "the central theoretical problem of sociobiology." Wilson states the problem in the following way: "How can altruism, which by definition reduces personal fitness, possibly evolve by natural selection?" In other words, why do animals sometimes behave in such a way as to benefit other members of their species when, by doing so, they may expose themselves to some danger?

Do they do it for the benefit of the species? Of course not. As we have seen, animals will often engage in behavior that is detrimental to their species, committing infanticide for example, if their personal fitness is thereby enhanced. Since the time of Darwin, the existence of altruistic behavior has been a difficult problem.

Perhaps the sociobiological explanation of this kind of behavior can best be understood by means of yet another example. Many different kinds of animals will give alarm calls when they become aware of approaching predators. For example, a prairie dog will cry out when it sees that a coyote or a hawk is approaching. Giving the call obviously benefits the other prairie dogs; they immediately run for the safety of their burrows. But it seems to lessen the probability that the animal giving the call will survive. After all, it has called attention to itself and has thus increased the chance that it will be eaten.

Since a prairie dog that runs silently to its burrow when it sees a coyote coming is a little more likely to survive than one which gives the alarm, we might expect that the genes that cause it to give a call would rapidly disappear from the population. After all, the noncalling animals would live longer and would produce more offspring. But this is obviously not what has happened. The existence of prairie dog "altruism" requires an explanation.

The paradox can be resolved if we consider the fact that every animal—whether it is a prairie dog, a duck or a human being—shares its genes with a number of other individuals. With the exception of certain social insects, every sexually reproducing animal receives half its genes from its father and half from its mother. Parents and offspring, then, share 50 percent of their genes with one another. So do siblings. If I have three brothers and one sister, one-half of the genes carried by each of them will also be present within my own body.

When the relationship is more distant, the percentage is smaller. I have 25 percent of the genes that were present in each of my grandparents, and I share 12½ percent with each of my cousins. It is this fact that allows sociobiologists to explain why the prairie dog behaves the way it does.

Any individual that gives up its life so that, say, three of its brothers may survive is actually enhancing its own fitness. When it dies, 100 percent of its genes are lost to the next generation. However, since 150 percent of its genes (the figure can be greater than 100 percent because some are present in more than one brother) are present in its three siblings, there is a net 50 percent gain.

Since a prairie dog is likely to be related to many other individuals in its "town," it is obviously acting in its own interest when it gives an alarm. For every child, parent or sibling that it saves, there is a net 50 percent gain; for every grandchild, 25 percent; for every aunt, uncle, niece or nephew, 25 percent; for every first cousin, 12½ percent. In most cases, it will be able to warn its relatives and escape with its own life as well. After all, giving the call does not ensure that it will die; it only enhances the probability of that unhappy fate.

This is the meaning of the term *inclusive fitness*. An animal maximizes its *personal* fitness by producing the largest possible number of successful offspring. It maximizes its *inclusive* fitness by behaving in such a way that the largest possible number of its genes are passed on to succeeding generations. Obviously, the former is nothing more than a special case of the latter. One doesn't always have to behave altruistically to pass along genes; it can be done by producing offspring of one's own.

The idea that animals maximize their inclusive fitness is called the *kin-selection theory*. This theory is not original with Wilson or any of the contemporary sociobiologists. It was first proposed more than forty years ago by the English geneticist J. B. S. Haldane. In *The Causes of Evolution*, published in 1932, Haldane wrote that "Insofar as it makes for the survival of one's descendants and near relations, altruistic behavior is a kind of Darwinian fitness, and may be expected to spread as the result of natural selection." On another occasion, Haldane expressed the idea even more succinctly. Asked, during a conversation in a pub, if he would give up his life for his brother, Haldane replied that he wouldn't. But he would gladly sacrifice himself for two brothers or for eight cousins, he added.

Obviously, one does not have to invoke the kin-selection theory to explain most kinds of animal behavior. When a seal fights for possession of a breeding territory, or when a baboon monopolizes an estrous female, it is doing nothing to enhance the fitness of its relatives. The same is true of most behavior that animals engage in. What the kin-selection theory does is explain certain kinds of behavior that would otherwise be very puzzling. Not only does it explain alarm calls in prairie dogs and birds, but also it explains why parents will defend their offspring at the risk of injury to themselves, and why a wild turkey will court females on behalf of its more dominant brother when it does not mate with the females itself. And, according to Wilson, kin selection may even explain the existence of homosexual behavior in human beings.

Some rather ambiguous evidence seems to indicate that a predisposition toward homosexuality may have a genetic component. This has led Wilson to suggest that there may be such a thing as "homosexual genes" and that kin selection may explain their persistence in the human species. The argument runs as follows: Homosexual members of primitive societies may have functioned as "helpers" who behaved in such a way as to enhance the fitness of close relatives. If they did, kin selection would ensure that genes favoring homosexuality would be passed on to future generations. Some of the "homosexual genes" would presumably be present in the relatives, even though they exhibited heterosexual behavior.

Naturally, many scientists are skeptical of this kind of argument. They point out that, even if homosexuality does have a genetic component, it does not follow that there are such things as homosexual genes. Certain combinations of genes might possibly predispose people toward homosexuality and toward any number of different kinds of behavior as well. In their view, to say that an individual's heredity can cause him to become a homosexual smacks of "genetic determinism."

Perhaps another example will make this point a little clearer. It is known that genetic factors are involved in a predisposition toward alcoholism. Nevertheless, modern scientists do not believe, as the eugenists did, that there is any such thing as a gene "for" alcoholism. Environmental factors are at least as important. They say that genes do not "make" an individual into an alcoholic in the same way that genes will cause a prairie dog to give an alarm call or a mallard drake to rape his mate.

Wilson's suggestions concerning genetic determinants of human behavior sparked the controversy about sociobiology that continues unabated today. Few biologists will doubt that genetic factors are an important determinant of numerous kinds of animal behavior. But many of them do not agree with the contention that such concepts are equally applicable to human beings.

When scientists began to perceive flaws in social Darwinism and in the assumptions that were made by the eugenists, a certain kind of conventional scientific wisdom began to develop. It began to be held that genetic factors played a relatively unimportant role in human behavior. Man's highly developed brain, it was maintained, made him almost infinitely adaptable. Furthermore, it was pointed out, man was the only animal that had culture. Culture allowed him to transcend the limitations of his genetic endowment.

As decades passed, the behavioral sciences made one discovery after another, and the findings seemed to reinforce this outlook. The work of the behaviorist psychologists caused the concept of human "instincts" to fall into disrepute and demonstrated how important learning and conditioning could be. Sociologists researched the effects of environmental influences. Anthropological studies like the ones that Mead conducted in New Guinea showed that cultural influences could shape human "nature" in surprising ways. The reactions against racism and aberrations like Nazi "anthropology" caused hereditarian outlooks to fall even more deeply into disrepute.

When popular books like Ardrey's *African Genesis* and Desmond Morris's *The Naked Ape* began to be published during the

1960s, they found a large audience. Many scientists, however, remained skeptical about the points that these authors were trying to make. Ardrey, Morris and authors who wrote in a similar vein assumed that there were important hereditary components in human behavior, and they gave evolutionary explanations that seemed to show how these components might have evolved. However, their arguments depended upon reconstructions of the evolutionary past and could not be proved.

When Morris discussed the development of the pair bond in prehuman hominids, he suggested that women had evolved fleshy breasts and buttocks so that they would be sexually more attractive and thus able to better cement their relationships with men. It all sounded reasonable enough. However, since there was no way of being certain that *Homo erectus* reacted to large breasts in the same way that American men* do today, it wasn't necessary to take the theory very seriously if one felt inclined not to. One could take the position that Morris hadn't given a verifiable explanation of the sexual preoccupations of contemporary males at all.

And then, suddenly, everything changed. In 1975, Wilson published his massive book on sociobiology. He demonstrated that sociobiological concepts could be successfully applied to numerous kinds of animal behavior. In the last chapter, he discussed the possible application of sociobiology to human beings. Wilson's theory worked so well that it had to be taken seriously. Scientists who were skeptical of hereditarian outlooks suddenly found themselves in the position of having to show why they thought that genetic concepts should not be applied to the study of human behavior.

As it turned out, Wilson's opponents were able to come up with a number of different arguments. Some of these have already been touched upon. One had to do with human culture. Anthropologists pointed out that human behavior was molded by culture, and they argued that cultural traits could not possibly be genetically deter-

*Or perhaps I should say "English-speaking men," since Desmond Morris is British.

mined. Cultural traits, they pointed out, could be acquired or wiped out within the space of a single generation. Behavior patterns which changed so quickly could not possibly have genetic components, they said. In human beings, they went on, the extremely slow process of shaping behavior by natural selection had been replaced by the much more rapid one of cultural evolution.

Other critics pointed out what they felt were methodological errors in sociobiology. Sociobiology explained nothing, they said, because it attempted to explain too much. They pointed out that it was possible to invent a genetic explanation for practically anything, provided that one was clever enough. Not only did sociobiology purport to explain a supposed male predilection for having a variety of sexual partners, it also claimed to be able to explain homosexuality. Men who were unfaithful to their mates were only following the dictates of genes that urged them to have as many offspring as possible. On the other hand, homosexuals—who might have no offspring at all—were able to pass their genes along by the process of kin selection. In the eyes of these critics, sociobiology seemed capable of explaining any kind of behavior. A theory that could do that, they said, was no theory at all, for it could neither be confirmed nor disproved.

Other scientists attacked sociobiology on genetic grounds. They pointed out that no gene that had a significant influence on human behavior had ever been discovered. When they discussed genetic control of human behavior, the sociobiologists were talking about something that had not yet been shown to exist. Admittedly, genes that caused retardation had been pinpointed. However, no one had ever discovered any genes for such things as altruism, spite, creativity and conformism, to mention only a few of the qualities that sociobiologists talked about. To speak of such things was to repeat the mistakes made by the eugenists, social Darwinists and criminal anthropologists during the late-nineteenth and early-twentieth centuries.

Although these arguments sound convincing, I think that we have to admit that they are more suggestive than conclusive. They don't demonstrate that sociobiology is false, only that its claims have not been proved. The sociobiologists, after all, do not say that all human behavior is under genetic control; they claim only that some of it is. Furthermore, the sociobiologists do not dispute the importance of human culture. They simply say that genetically determined tendencies limit the forms that culture can take. Wilson expresses this idea in the following way: "Although the genes have given up most of their sovereignty, they maintain a certain amount of influence in at least the behavioral qualities that underlie variations between cultures." Or, as he has also written, genes "hold culture on a leash."

If only more were known about the effects of genetic DNA on human behavior, the controversy might be resolved. Unfortunately, our knowledge in this area is very limited. Although more than a thousand human genes have been identified, and several hundred have been mapped on specific chromosomes, the function of the vast majority of the genetic material contained in the human body is unknown. Scientists do not even know how many genes man has. It is sometimes stated that each of us has approximately a hundred thousand. But this figure is only an estimate. The true figure could be even higher.

Geneticists do not yet understand why most of the DNA that is found in human cells even exists. A small percentage—between 0.5 and 5 percent—makes all of the body's proteins. No one has yet been able to explain what, if anything, the remaining 99.5 to 95 percent does. No one knows exactly how DNA synthesis of proteins is turned on and off, and the connection between protein synthesis and physiological and behavioral traits is not well understood. It is true that no one has discovered a gene for altruism. But then no one knows how genes go about making a human knee either.

It has been established that many human traits are hereditary. Eye color, blood type, and susceptibility to certain types of cancer are just a few. But knowing that a trait is genetically determined and identifying the genes that code for it are two different things. The fact that most human traits are determined by a number of different genes working together makes the situation all the more complicated.

So if the sociobiologists assume the existence of genes for certain behavior traits, they can't really be accused of doing sloppy science or of committing methodological errors. It is equally necessary to assume, without proof, that there must exist genes for such traits as blond hair.

The sociobiologists claim that some human behavior is under genetic control. I don't see how one can deny that. We have already considered some human traits—such as the formation of dominance hierarchies—that must have been shaped by natural selection. They are so commonly encountered in the animal world that the only reasonable assumption is that they appeared in our prehuman ancestors long before we evolved into *Homo sapiens*. There may be specifically human traits—inhibitions against killing the infants of strange females, for example—that also evolved by natural selection. There are probably many others as well. After all, natural selection did not stop acting on human behaviorial patterns the moment that culture was invented.

Nevertheless, I cannot help but be skeptical about some of the claims that the sociobiologists make. The assumption that there must be genes for altruism in human beings, in particular, seems to me to be very questionable.

It seems clear that genetic factors cause prairie dogs to give alarm calls. The genes that do this cannot be identified, and it is not known whether one gene or many genes code for this behavior. But, since there is no other reasonable explanation for this behavior, the idea that there are genes for alarm calls is easy to accept.

However, it is one thing to say that genetic factors cause prairie dogs to engage in a very specific kind of behavior, and quite another to say that there are genes that cause human beings to behave altruistically under a wide variety of circumstances. Giving an alarm when a coyote approaches and having generalized altruistic inclinations seem to me to be two very different things.

The sociobiologists admit that altruism in human beings can be expressed in complex ways. In fact, they postulate the existence of two kinds of altruism in the human species. Kin selection, they say, causes us to be nice to our relatives, while genes for *reciprocal altruism* cause us to be helpful to those to whom we are not related.

Reciprocal altruism, the sociobiologists say, is really not altruism at all. On the contrary, it is nothing more than selfishness in disguise. We behave altruistically toward nonrelatives because we expect something in return. We do favors for other people because we expect that they will do the same for us.

I can't avoid thinking that the sociobiologists are on shaky ground here. They find it necessary to postulate the existence of not one, but two kinds of altruism in human beings. And yet, neither seems to have very much to do with the genetically determined "altruism" whose existence is so well established in animals. The human equivalent of genes for alarm calls in prairie dogs, after all, would be genes that coded specifically for some such behavior as saving other human beings from drowning.

Sociobiology seems to be successful enough when it is applied to animals. However, when one tries to extend it to human beings, it seems to be necessary to go through various sorts of contortions. A theory that "human beings are genetically predisposed to behave altruistically, except when they are really being selfish" hardly seems to be adequate. A successful theory of human behavior ought to be more elegant than that. It should not be necessary to introduce this kind of additional assumption just to make it work.

Some of the controversy that has surrounded sociobiology has been political in nature. Some scientists have argued against the assumptions of the sociobiologists on political grounds. They say that sociobiology can be used to justify racism and the continuance of sexual inequality. Those who lean toward the political left have sometimes even gone further, charging that sociobiology seeks to justify the political status quo by making it seem "natural."

There seems to be some substance to at least some of these charges. Right-wing extremists in Britain and France have attempted to use sociobiological ideas to justify their outlooks. In Britain, for example, the right-wing journal *New Nation* has characterized the opponents of sociobiology as "Marxist" and "Jewish," and has informed its readers that

> for us, as racial nationalists, [sociobiology] is an important vindication of our position. . . . what the evolutionary theoreticians have shown us is that, with the system of genetic inheritance shared by all vertebrates, the only kind of social organization which can evolve, let alone work, is one based upon kinship, upon the ties of blood and race.

For the benefit of those who are unaware of the fact, I should perhaps mention that "blood and race" was one of the preoccupations of Hitler.

In the United States, articles about sociobiology have suggested that there may be fundamental biological reasons why women will not achieve full social equality. For example, one such story in *The New York Times* (Nov. 30, 1977) stated that "even some staunch feminists are reluctantly reaching the conclusion that women's aspirations may ultimately be limited by inherent biological differences that will forever leave men the dominant sex."

The sociobiologists have been careful to dissociate themselves from those who would distort their theories in order to justify rightist political and radical doctrines. However, they often seem to endorse the idea that behavioral differences between men and women are not cultural, but innate. For example, Wilson suggests that male dominance is an innate biological trait in human beings.

The idea that limits should be placed on sexual equality, and the idea that there are innate behavioral differences between men and women are, of course, two different things. In fact the latter is endorsed by some feminists (they think that women are superior to men). However the idea that male dominance is natural, that men are more competitive, while women are better suited to "nurturing" roles is hardly conducive to social or economic equality between the sexes.

Wilson, it is true, doesn't express the latter idea in quite so blatant a way. But he does state that "women as a group are less assertive and physically aggressive" than men. Elsewhere, he states that girls are not as "venturesome" as boys. Furthermore, he claims that there is evidence that would indicate that these differences are hereditary.

Those of us who wish to disagree with such conclusions cannot argue that sociobiology must therefore be false. It is one thing to present evidence that indicates that the sociobiologists are wrong, and quite another to say that we would like to believe that they are mistaken. But, in the meantime, since the correctness of sociobiological ideas has not been proved, we are certainly within our rights if we accuse Wilson of making unsubstantiated statements that sound very much like propaganda for male dominance.

Wilson has suggested that differences between human cultures may be partly genetic. If sociobiologists ever are able to present evidence that this is true, it is practically certain that some will refuse to accept the idea on the ground that it is "racist." In this case, however, it is a little harder to criticize them for their suggestions. We have become sophisticated enough to reject the racial and ethnic stereotypes that have been so common in the past. But it doesn't necessarily follow that there are no differences at all. There is no evidence that any racial or ethnic group is inferior to any other in intelligence, in creativity, or even in "altruism." However, cultural differences are obviously very real. At present there is no way of knowing whether genetic factors play any role. It is certainly possible.

Although there are reasons for maintaining a skeptical attitude, it would probably be best to retain an open mind about the validity of the sociobiological enterprise until such time as more is understood about human genetics. We may find that we have biases that incline us either to accept sociobiological ideas or to repudiate them. If we do, we should be careful not to let these biases influence us too strongly. As we have seen, preconceptions and political affiliation have influenced thought about human nature numerous times. In some cases, they have even led to incorrect interpretations of "hard" facts. There is every reason to think that human predispositions will be just as influential in the future.

There is one particular kind of predisposition that has cropped up again and again. Environmentalism has been associated with liberal and radical political outlooks, while hereditarian interpretations have been bound up with conservative viewpoints. Although we would like to believe that science and politics should be divorced from each other, they frequently have not been.

It should come as no surprise that sociobiology, a theory (or discipline) that embraces hereditarian outlooks, should have aroused the ire of the left. Some of the most vehement criticisms that have been heaped upon Wilson and his colleagues have come from a group affiliated with the radically oriented Science for the People. This group, a collective calling itself the Sociobiology Study Group, was made up primarily of Harvard professors and students. It began to publish attacks on Wilson's *Sociobiology* shortly after the book was published. Although some of the charges were reasoned scientific objections, others were unabashedly political in nature.

In a statement published in *The New York Review of Books*, the Sociobiology Study Group characterized Wilson's book as an attempt to reinvigorate theories that had, in the past, "provided an important basis for the enactment of sterilization laws and restrictive immigration laws by the United States between 1910 and 1930 and also for the eugenics policies which led to the establishment of gas chambers in Nazi Germany." The Group charged further that Wilson had joined "the long parade of biological determinists whose work had served to buttress the institutions of their society by exonerating them from responsibility for social problems."

Wilson responded by denying that Sociobiology contained any political message. He complained that the Study Group had been guilty of misrepresenting his viewpoints. He charged, furthermore, that the activities of the Group had incited a campaign of intimidation. Wilson pointed out that, after the Study Group's attack had been published, student members of the Harvard-Radcliffe Committee against Racism had distributed through the Boston area leaflets that labeled the book as "dangerously racist." Wilson pointed out, also, that he had had to face hostile questioning when he gave lectures.

The assumption that Wilson is an unwitting tool of the ruling class and that *Sociobiology: The New Synthesis* contains implicit political messages is debatable, to say the least. Wilson is a scientist who was simply trying to point the way to a better understanding of human nature. Although he does seem to have a few antifeminist biases, *Sociobiology* is hardly a neo-Nazi tract. Wilson's only crime was taking a hereditarian point of view, an outlook that has been associated with conservative viewpoints in the past. It is conceivable that Wilson might have some conservative sympathies (although he describes himself as a liberal), but if he does, they are not discernible in his book.

Although there were no political messages in *Sociobiology*, the same cannot be said of Wilson's more recent books. Possibly as a response to criticism from the left, he has included short critiques of Marxism in two works that he has published since the controversy began. In *On Human Nature*, he says that Marxism is "sociobiology without biology," and he charges that "the strongest opposition to the scientific study of human nature has come from a small group of Marxist biologists and anthropologists who are committed to the view that human behavior arises from a very few unstructured drives." Calling Marxism "an inaccurate product of scientific materialism," he claims that it is now "mortally threatened by the discoveries of human sociobiology." In *Genes, Mind and Culture*, a book that he wrote with physicist Charles J. Lumsden, Wilson repeats the charges in a somewhat more restrained manner.

It is probably safe to say that Wilson has no more demolished Marxist political philosophy than the Sociobiology Study Group did away with sociobiology. However, the charges made on both sides demonstrate once again that science and politics are likely to find themselves entangled with each other whenever they attempt to draw conclusions about human nature. Since the time of Locke, such theories have been used for political purposes time and time again, while scientists have been influenced by political predilections.

Perhaps it would be going a bit too far to say that science and politics must inevitably find themselves entangled in this manner. But it happens often enough to warrant the prediction that debates such as the one that erupted over sociobiology will inevitably happen again.

10

The Nature of Human Nature

At this point, one might be tempted to ask whether anything at all has been learned about human nature. We have seen that theories about human nature can be used to serve political ends. We have observed that prejudices, predilections and political leanings have had an enormous influence on scientific thought on this topic. But have scientists learned anything about the nature of human nature? Can we even be sure that there is any such thing as a human nature that is shared by people in different cultures?

I think that it is clear that both these questions can be answered with an unqualified Yes. In spite of all the difficulties, errors, misinterpretations, blind alleys and political entanglements, man has discovered some important facts about himself during the last century or so. To be sure, there have been a lot of mistakes and a lot of time-consuming detours. However, there are certain things that can be said about human nature with a great amount of confidence.

It is obvious that culture influences our behavior to a much greater extent than our genetic endowment does. Nevertheless, innate behaviorial traits do exist. Many such patterns can be observed in infants. Crying is one such pattern. So is smiling. A baby will began to smile when it is about seven weeks old. It is known that this behavior is not the result of learning, because infants that are blind from birth exhibit it also. One is forced to conclude that a predisposition for smiling is innate.

Startle responses to noise are also programmed into infants. So is the behavior that is associated with nursing. Babies do not have to be taught to seek the breast, to suck on a nipple, or to swallow. Newborn babies also have a grasping reflex. A touch on the palm or on the soles of the feet induces a grasping response that is so strong that an infant is able to support its own weight in a clinging position shortly after birth.

Infants begin to show anger when they are about four months old. Full temper tantrums develop during the second year after birth. It is obvious that this is an innate behavior pattern too. Tantrums are not something that a baby can learn by imitating its parents.

Certain kinds of learning capacities also seem to be innate. When an infant reaches a certain age, it will learn to crawl. Later on, its coordination will have increased to such a point that it will be able to learn how to walk. The acquisition of language seems also to be governed by innate programs. If this were not the case, it would not be so easy for a child to learn a language and so difficult for an adult. Young children are perfectly capable of learning two languages simultaneously (for example, when their parents are bilingual), and they have no difficulty with the pronunciation of either. An adult, on the other hand, will rarely learn to speak a language without any trace of an accent.

It is more difficult to discern innate behavior patterns in adults. But this does not mean that they are not present, only that they are modified by learning. Furthermore, it is not always possible to determine what is learned and what is innate. The controversies about altruistic behavior provide a good example of this. There might be genes for altruism, but altruistic behavior could equally well be learned. No society could continue to exist if it did not encourage a certain amount of cooperation among its members. Hence altruism may be nothing more than a culturally conditioned trait that encourages cooperative behavior. Even if altruism is innate, the expression of it is certainly affected by learning. Margaret Mead did not encounter many altruists among the Mundugumor.

On the other hand, the human emotions are certainly innate. Human cultures teach people to become angry about particular things, but the emotion itself does not seem to be something that has to be learned. When we become angry, certain physiological changes—changes that are not under conscious control—take place within the body. Adrenaline is released into the bloodstream. The heart begins to beat more rapidly, and blood is transferred from the skin and viscera to the muscles and the brain. Concentrations of blood sugar increase. Digestion stops. Breathing becomes more rapid. And yet an angry person is unaware that these changes are taking place. This in itself is evidence that innate programs are operating.

The physiological changes that are associated with other emotional states are sometimes less dramatic. However, these states often elicit gestures and behavior patterns that make them easy to observe. Smiling, frowning, laughing and weeping are found throughout the human species. To a certain extent, these patterns are under conscious control, and cultural learning certainly modifies the modes of expression. For example, some cultures encourage women to be "emotional" and discourage the same behavior in men. But there is no evidence that this has any effect on our subjective feelings. In some societies, men are not supposed to be seen weeping. Nevertheless, they still experience sorrow. The capacity to do that is innate.

It is easy to understand why some emotions have evolved. Anger and fear, for example, prepare the body for aggression or for flight. Others are more puzzling. What is the function of grief? Could it be that it exists to bind the members of a social group more closely together? Does it exist for some other reason? Whatever its cause, there is no doubt that it is innate. The expression of emotion is shaped by the human brain, specifically by the hypothalamus and the limbic system. And the human brain is something that evolved by natural selection.

Man is a social animal. Like the chimpanzee, the dolphin, the baboon, we congregate in groups. There is every reason to believe that our need to interact with other human beings is innate too. Hermits exist, but they are rare in every society.

It is true that some individuals are more gregarious than others. But this could be the result of learning. If a behavioral trait, such as gregariousness, is "wired in" to our brains, it does not follow that this trait cannot be modified by environmental influences. In fact, we should expect that learning should modify most innate behavior to some extent. This, by the way, is what makes it so difficult to tell whether or not differences between human beings are genetic. It is obvious that some people become angry more readily than others, for example, but there seems to be no way of telling whether they do this because their genetic endowment is different or because they have been brought up differently.

But surely there is nothing very earth-shattering about the discovery that human beings experience love, hate, anger, guilt and sorrow, and that the expression of these emotions is modified by experience? This fact was recognized thousands of years ago, at the very least. Has nothing more been learned about human nature?

I think that it is possible to say that quite a bit has been learned, because we have been able to profit from our mistakes. It is precisely because science has allowed itself to be led astray so many times that our knowledge has so dramatically increased.

The Danish physicist Niels Bohr once defined *expert* as "a man who has made all the mistakes." We have certainly made enough mistakes. It is impossible to doubt that we have profited enormously from them. The Dart-Adrey killer-ape hypothesis provides a good example of this. It is clear now that this theory of innate aggression is mistaken. But this does not mean that the debates over it were a complete waste of time. On the contrary, the theory prompted scientists to probe more deeply into questions related to human aggressiveness. New observations were made, experiments were performed, and the subject was discussed endlessly in books and in articles that were published in scientific journals. As a result, science has a much better understanding of the nature of aggression than it ever did before.

Although controversies about human aggression still go on, a number of facts have become clear. It has been discovered that *Homo sapiens* is not the only species in which individuals will kill conspecifics. It has become apparent that, compared with certain other animal species, human beings are not especially aggressive at all. It is true that our technological capabilities have given us the ability to commit racial suicide. But matters could be much worse. We could be really aggressive animals, like the various species of baboons, for example. It has been suggested that if baboons possessed nuclear weapons, then the world would be destroyed within a week.

On the other hand, we have learned to be somewhat suspicious about sentiments concerning the "nobility" of human beings. Although the sociobiologists have not shown that altruism is produced by any particular gene or combination of genes, they have pointed out that altruistic behavior has a selfish component. An animal that behaves in such a way as to benefit its relatives is just enhancing its own biological fitness. And the concept of reciprocal altruism makes us more aware that the reason for our willingness to help nonrelatives may be that we expect other human beings to give us aid when we need it.

The mistakes that the early-twentieth-century anthropologists made about Neanderthal man are illuminating too. When they projected so many "bestial" characteristics upon Neanderthal, it was difficult to resist the implication that evolution had somehow brought us to a higher plane. By magnifying the difference between us and our immediate ancestors, they increased the apparent distance between us and the other animals.

Subsequent discoveries have shown us how mistaken this view was. We now realize that Neanderthal was very much like us. It has begun to seem increasingly likely that *Homo erectus* was probably not so different either. True, he had a slightly smaller brain than the one that we possess. But he did hunt and use fire. It seems likely that he possessed the use of language also.

Narrowing the distance between us and our ancestors has made it easier to see our kinship with other animal species, especially the primates. The realization that we share certain behavioral traits with our primate relatives has given us real insight into our own nature. For example, when we observe a dominant macaque that is kept in power by its allies, we cannot help but be struck by the similarity between this behavior and the functioning of certain human hierarchies.

Admittedly, comparisons between humans and animals can be misleading. In India, attempts are sometimes made to justify the caste system by appealing to the existence of castes in insect societies. Obviously the two different kinds of castes have little to do with each other. Similarly, the existence of baboon harems does not imply that polygamy is the most natural mating system for human beings. But there is nothing wrong with speculation. If the observation of very "human" behavior in animals leads us to new hypotheses, that is all we should ask. If we make mistakes, we should be able to learn from them, just as we have done in the past.

The most insidious errors have been those concerning the differences between human beings. Here too the making of mistakes has led to greater understanding. If no attempt had ever been made to find scientific justification for racism, we might not yet realize how small the differences between the human races really are. Similarly, we have learned to repudiate the idea that there are people who are "born criminals." We realize now that the inmates of prisons tend to be people with little education, who have had few opportunities in life. Although we are not willing to exonerate them for their antisocial acts, we do recognize that environmental factors have played an important role in steering them toward criminal acts. Sometimes the realization that a trait is not part of someone's "nature" is as important as finding traits that are hereditary.

Of course, some mistakes are made more than once. In recent years, the "born criminal" theory was revived in a modern form. It was discovered that men with double Y chromosomes constituted a disproportionately large number of those confined to prisons and institutions for the criminally insane. It was reasoned that since men normally have one X and one Y chromosome, while women have two X's, the abnormal presence of an extra Y in males might be likely to cause violent behavior.

Although the evidence seemed conclusive when it was first presented, doubts were soon cast on this interpretation. It was pointed out that most XYY men lived normal lives, and that those who had been convicted of crimes were no more violent than criminals in general. In fact, there was evidence that they were less so. It turned out that many of them had been convicted of minor offenses and had been given mild penalties.

If a larger percentage of XYY men went to jail, various investigators reasoned, the most probable cause was low intelligence. Like the victims of many other chromosomal abnormalities, many XYY males exhibited a certain amount of mental retardation. It had long been known that low intelligence and criminal behavior were correlated in normal men (probably because low intelligence is likely to lead to a poor education and low economic status).

Many of the questions that have been raised have not been answered. It has not been explained why men with an extra X chromosome (XXY) seem to commit fewer crimes (or to get caught less often) than those with an extra Y. However, the notion that there is a connection between the presence of an extra Y chromosome and the tendency toward violent behavior has not stood up to scrutiny, and the idea that there is such a thing as a genetic predisposition for committing criminal acts has once again been discarded.

No one would deny that there are differences between human beings and that those differences are partly genetic. However, it is apparent that the various theories of genetic determinism from the time of Galton to the present are almost certainly false. There is no such thing as hereditary genius that will cause one to rise to a position of eminence, whatever the obstacles. One might be illiterate, for example, or labor under the disability of growing up in a ghetto environment. Only a few of us have the capacities that—given the right environment—would enable us to become a Mozart or an Einstein. However, most of us have the ability to fill almost any social role, provided that we receive the right kind of training or upbringing.

Certainly some innate behavioral predispositions exist. It is possible to discern personality differences in infants that are only a few weeks old. However, culture, learning and environmental influences seem to play a more important role in the molding of human character. Our genetic inheritance might sometimes steer us in one direction or another, but it is not a straitjacket.

It is not even proved that hereditary components of I.Q. make very much difference. In the first place, it is known that environmental factors have an important effect on I.Q. scores. This, by the way, is why children born to parents of high socioeconomic status tend to do better on I.Q. tests. It also seems to be the reason for the difference between average white and average black I.Q.s. Furthermore, although I.Q. seems to be a good indicator of academic success, it does not correlate especially well with anything else. It may not even be a good measure of that quality that we colloquially refer to as "intelligence." As we all know, there are some very intelligent people who do not do well in school. Albert Einstein and the great Roman Catholic philosopher Thomas Aquinas are just two examples of individuals who, at one time, were thought to be "slow."

An erroneous theory with far more drastic implications than Galton's notion is the belief of the eugenists that there are classes of genetically inferior individuals who will degrade the human race if they are allowed to continue to breed. This belief is closely related to the theory that the more intelligent members of our society produce fewer offspring. At one time many scientists feared that the average intelligence of the American population would gradually decline. Indeed, decades ago, published statistics seemed to indicate that the people with the lowest I.Q.'s had the largest number of babies.

It is now known that the fear of a declining average intelligence was based on two errors. In the first place, the correlation between I.Q. and birth rate merely reflected the fact that people in the higher socioeconomic classes have smaller families, and since these people generally have had more favorable environments, their I.Q.'s tend to be higher. The statistic does not necessarily signify that there is any connection between the number of children that one has and one's innate intelligence.

Secondly, the statistic itself was incorrect. People with high I.Q.'s have a *higher* average number of offspring. It is true that high-I.Q. couples tend not to have a lot of children. But this is more than offset by the fact that people with low I.Q.'s are less likely to get married in the first place.*

Questions about the genetic component of differences between individuals are certainly interesting. And they are certainly part of our subject, at least in the broad way that I have defined it. Nevertheless, they sometimes seem to be of peripheral importance compared to the question "What is basic human nature?" This is the question that causes us to exercise our imaginations to the utmost. It is the question that has been contemplated by philosophers over the ages. It is the question that has most concerned scientists ever since they began to suspect that a science of human nature might be possible.

Even today, we know surprisingly little about what this "basic" human nature might be. We all think that we understand it, but when we subject our ideas on the subject to closer scrutiny, we realize that we view our fellow human beings in the ways that we have been taught. Every culture promulgates certain ideas about what people are like, and every culture is different. This implies that culturally ingrained ideas on the subject should be considered especially suspect. In the earlier chapters of this book, we saw what problems Victorian science caused itself by remaining obliv-

*Of course they could still have illegitimate offspring, so one can't be absolutely sure that their fertility is lower. Nevertheless, the assumption of a lower fertility rate seems a good guess.

ious to its own ethnocentrism. What is human nature? An American would give one answer, an Arapesh an entirely different one, and a Mundugumor yet a third.

One might object that many different cultures have now been studied by anthropologists. Shouldn't it be possible to pick out human behavior traits that are universal? Perhaps. Edward O. Wilson gives such a list. He lists seven cultural items that, he says, are found throughout the world. These are property rights, body adornment, incest taboos, sexual roles, rites of passage, intraspecific war, and belief in the supernatural.

It is possible to quibble with this list to some extent. For example, not all peoples engage in war. And when we look at these items closely, another problem crops up. At present it is not possible to tell to what extent these practices are part of our genetic inheritance. Some of them might very well have been invented independently by cultures around the world. Furthermore, it is difficult to tell what basic human traits are responsible for these practices. Take incest taboos, for example. Do they exist because we have a horror of incest or because we sometimes find the idea tempting? Is belief in the supernatural an innate predisposition, or does it stem from a human need to explain natural phenomena?

At this point, it might not be inappropriate to engage in a little speculation. If universal human propensities exist, then it should be possible to point out what some of them are. It may not be possible to explain how they evolved, or to say to what extent they are under genetic control. As we have observed, many human traits are the result of cultural, rather than biological, evolution. Even traits that are universal may be affected more by culture than by biology.

But we can certainly leave these problems to future generations, which will presumably have a better understanding of behavioral genetics. If we construct a few plausible hypotheses, that should be enough for now. Of course, it must be remembered that they are hypotheses, and that even very plausible ideas sometimes turn out to be wrong. After all, we are just as influenced by prejudices and predispositions as our predecessors were. We certainly have a tendency to believe what we want to believe, just as they did.

Keeping those reservations in mind, let us see what we can come up with.

Human beings are cooperative. It would be hard to believe that this is not a genetically controlled trait. After all, as has been pointed out many times previously, man is a social animal. Social interactions would hardly be possible if there was not a genetic predisposition toward cooperation.

Human beings are passionate. Much has been made of the size and complexity of the human brain. It is often implied that it is our intellectual capacity alone that distinguishes us from the other animals. Certainly, intellect plays an important role. But is it the only quality that distinguishes us from the other species?

I tend to think that it isn't. In fact, I would go so far as to characterize human beings not as thinking animals, but as animals that take a passionate interest in things. Without intellectual passions, there would be no human science, no art, no literature. There would be no such things as philosophy or mathematics or religion.

If they were lacking in passion, people would not argue about politics. They would not interest themselves in the fortunes of the local baseball or football team. They would not become passionately interested in collecting stamps or in playing backgammon or in programming computers.

Intellectual passion sometimes exhibits itself in surprising ways. For example, chess masters will sometimes spend weeks analyzing positions from world championship games in order to learn the "truth" about this or that encounter. We really are a strange kind of animal if some of us will labor so hard looking for the "truth" in chess, which, after all, is only a game.

Members of primitive cultures are as passionate as individuals who live in technologically advanced societies. The human beings who live in those few hunting-gathering societies that manage to survive are generally excellent naturalists. They will often spend hours doing nothing but observing the behavior of animals around them. As a result, their knowledge of animal behavior often equals or surpasses that of Western scientists.

The capacities of the human brain are indeed amazing. It is especially surprising that a brain that evolved in order to make it easier for our ancestors to survive on the African savannah should be able to comprehend such things as complicated theorems in mathematics. Nevertheless, intellectual capacity would mean very little if we did not possess a drive that induced us to take a passionate interest in things. Since intellectual passion, like sexual passion, is an emotion, there is little doubt that the capacity to experience it is innate.

Human beings participate in collective belief systems. Some of the critics of sociobiology have vehemently attacked certain statements that Wilson has made about human indoctrinability. They have objected to his statements to the effect that "human beings are absurdly easy to indoctrinate," and that "men would rather believe than know." They have heaped scorn upon his suggestion that there might be genes for conformity.

In this case, I can't help thinking that it is Wilson who is more nearly correct. Every human culture has its belief systems, and every culture ostracizes those who do not accept socially sanctioned beliefs. It is not possible to find a human society whose members are not expected to have unwavering faith in some set of "truths."

Sometimes these beliefs are political in nature, and sometimes they are nonpolitical. Sometimes they are beliefs about human nature. In certain parts of the world, one encounters beliefs about the spirits of one's ancestors. In others, people believe in voodoo. In certain cultures, masculinity is equated with *machismo*. In others, it is thought that older women are especially attractive sexually. In yet others, it is thought that the human body must be bathed and deodorized constantly, and that mouthwash can increase one's sex appeal.

Deviants who do not accept these socially sanctioned beliefs are not easily tolerated. Human beings have proved quite willing to torture witches, burn heretics, and to hang, crucify and slaughter religious and political dissidents. They react with irritation or rage even to minor offenses. For example, in our culture, failure to bathe frequently (it should really be a matter of personal choice, shouldn't it?) results in social ostracism, and the individual who does not dress according to the dictates of social convention will not be able to negotiate job interviews successfully.

In various societies it is required that one profess belief in a bewilderingly large number of different, sometimes mutually contradictory, ideals. In medieval Christendom, one was required to believe that human beings were naturally sinful. In the contemporary United States it is necessary to profess belief in the importance of civil liberties, even when one is attempting to limit them. In Communist states, one is required to believe in the supremacy of the proletariat, even when the workers are having someone else's will imposed upon them from above.

The sanctions that are imposed upon deviants can be severe. However, they do not have to be imposed very often. Most individuals want to accept the belief systems that are promulgated by their cultures. They have a desire to accept the values of the people around them. One of the things that children fear most is being thought different. Adults are sometimes a little less fearful in this respect, but they are just as avid about attempting to conform.

Conformity in belief can extend even to ideas that are known to be suspect. The current preoccupation with astrology provides a good example of this. Even people who understand very well that astrology has no rational foundation find themselves being caught up in discussions of the supposed characteristics of individuals who have been born under any given zodiacal sign. It is perfectly possible to believe intellectually that astrology is nonsense while discussing horoscopes with one's neighbors. The activity is socially sanctioned, and that is all that matters.

Human beings distrust strangers. This is probably related to the trait that has just been discussed. The trouble with strangers is that they adhere to belief systems that are at variance with our own. The members of many primitive societies do not even consider neighboring tribes to be fully human. In fact, their names for themselves can often be translated more or less accurately as "human beings." The implication, of course, is that *we* do and believe those things that it is natural for human beings to believe. "They" do not.

Human beings look for justice in a world that contains none. I can't help but think that it is this, rather than the desire for life after death, that is the motivating factor in religious belief. After all, belief in a future life is not a characteristic of all religions. It played no role in ancient Judaism, for example. The Buddha refused even to discuss questions concerning life after death; he maintained that one should be concerned with one's problems on earth.

Religions do promise justice, however. Christianity answered the question "Why do the wicked often prosper?" by promising that wicked individuals would be punished after death, while the good would be granted eternal life. (During medieval times, by the way, one of the delights of heaven was thought to consist of the opportunity of seeing one's enemies burn in hell.) In certain Eastern religions, the doctrine of karma requires that one must spend future lives making restitution for the mistakes that are made in the present one.

The striving for justice motivates politics as well. Different political systems and conflicts between the proponents of different political ideals exist because we do not all have the same conception of what a just society would be like. The desire for justice may even have motivated the various scientific justifications of racism that were once so common. The Victorian scientists, after all, had little personal interest in the oppression of colonial peoples. But they wanted very much to believe that their societies were not treating members of other races in an unjust way. So they attempted to prove that racial inequalities existed. Today, sentiments such as those expressed in Kipling's poem "The White Man's Burden" seem ludicrous. But in Kipling's time they had a strong appeal to people who sincerely wanted to believe that they were behaving altruistically.

Similarly, social-Darwinist theories were an attempt to show that late-nineteenth-century capitalist society was not as unjust as it seemed to be. Applying the idea of natural selection to human society acted to assuage any guilt that the more privileged might otherwise have experienced. By arguing that it was a "scientific fact" that the lower rungs of society should be occupied by the least able, by promulgating the myth that the existence of extreme poverty was a natural consequence of the "struggle for existence," our forefathers were able to convince themselves that inequality was part of the natural order of things.

And what is the human emotion of envy, after all, but the ire aroused by the feeling that someone has gotten something that he does not deserve? We do not become automatically envious of those who are better off than we are. We experience that emotion only when we believe that someone else's good fortune, or our own bad luck, is unjust. Few of us envy the rich and the famous; many of us envy competitors whose success we feel to be undeserved.

Human beings have dominance hierarchies. This topic has been discussed. However, a few additional observations can be made. I think that it is safe to say that most of us realize that human beings can be very competitive. But we sometimes lose sight of the fact that this is not always the case.

Many people, once they have attained a certain social status, attempt to rise no higher. It is one of our characteristics that we are not always competing to rise to a higher level in this or that hierarchy. In fact, most individuals will placidly accept whatever status they have attained by the time they are thirty or forty. Few people change occupations at this age, and even those who have no families rarely resume their education.

People will struggle fiercely to prevent any lowering of their status. But after they reach a certain age, most of them lose their drive to struggle upward. It is typically the young who are venturesome. Older people tend to be more restrained.

If this was not the case, society would not function as smoothly as it does. Dominance hierarchies would have little meaning if each individual were constantly trying to disrupt them. As it is, technological societies have problems enough with crime, with pollution, with trying to guarantee racial and ethnic equality, with trying to smooth out the struggles between various political and economic interests. It is difficult to imagine what things would be like if the hierarchal structures that we have developed to deal with such problems were less stable than they are.

Like us, chimpanzees struggle with one another for dominance. However, it is the younger individuals who challenge the older ones. Furthermore, most individuals never attempt to attain leadership of a social group. The vast majority seem to be perfectly content with having attained a place somewhere in the middle. Human beings behave in the same way. Only a few possess a burning drive to rise to the top of their profession. Most are more comfortable with some modest kind of success.

Lists like this can be extended indefinitely. It would certainly be possible to add a number of other common human characteristics. However, we might as well stop here. I have listed those human traits that seem to me to be the most striking. As I have pointed out, I have no way of knowing what genetic factors have brought them about. But I can't help believing that genetics plays a prominent role in all of them.

One could conceivably argue that some of the traits I have listed are primarily cultural. But I am not sure that such arguments would be entirely convincing. And I don't think that such arguments could be made about all of them.

Some of the traits have almost certainly been shaped by natural selection. For example, let us take one that was mentioned in the paragraphs on dominance, the venturesomeness of youth. Surely this is part of inherited human nature. After all, wouldn't most societies—and most parents—condition their young to behave in a slightly more sedate manner, if they possibly could? Isn't it obvious that young people persist in being the way they are because they have certain genetic characteristics that have been inherited from our ancestors on the African plains?

Bibliography

Aaron, Richard I. *John Locke*, 3d ed. Oxford: Clarendon, 1971.

Alland, Alexander, Jr. *Evolution and Human Behavior*, 2d ed. Garden City: Doubleday, Anchor, 1973.

ApSimon, A.M. "The Last Neanderthal in France?" *Nature* Vol. 287 (1980), pp. 271–72.

Ardrey, Robert. *African Genesis*. New York: Bantam, 1977.

Barash, David P. *Sociobiology and Behavior*. New York: Elsevier, 1977.

———. *The Whisperings Within*. New York: Harper & Row, 1979.

Barlow, George W., and Silverberg, James, eds. *Sociobiology: Beyond Nature/Nurture*. Boulder, CO: Westview, 1980.

Barzun, Jacques. *Darwin, Marx, Wagner*. Garden City: Doubleday, Anchor, 1958.

Blacker, C.P. *Eugenics: Galton and After*. Cambridge, MA: Harvard University Press, 1952.

Boule, Marcellin. *Fossil Men*. Edinburgh: Oliver and Boyd, 1923.

Brace, G. Loring. *The Stages of Human Evolution*, 2d ed. Englewood Cliffs: Prentice-Hall, 1979.

Caplan, Arthur L., ed. *The Sociobiology Debate*. New York: Harper & Row, 1978.

Cavalli-Sforza, L.L. *Elements of Human Genetics*, 2d ed. Menlo Park, CA: Benjamin, 1977.

Cloninger, C. Robert, and Yokoyama, Shozo. "The Channeling of Social Behavior." *Science* Vol. 213 (1981), pp. 749–51.

Cohen, David. *J.B. Watson*. London: Routledge & Kegan Paul, 1979.

Collins, Desmond. *The Human Revolution*. New York: Dutton, 1976.

Cranston, Maurice. *John Locke*. London: Longmans, 1957.

Cravens, Hamilton. *The Triumph of Evolution*. Philadelphia: University of Pennsylvania Press, 1978.

Cronin, J.E., et al. "Tempo and Mode in Hominid Evolution." *Nature* Vol. 292 (1981), pp. 113–22.

Curti, Merle. *Human Nature in American Thought*. Madison, WI: University of Wisconsin Press, 1980.

Darwin, Charles. *The Descent of Man*. Princeton, NJ: Princeton University Press, 1981.

———. *The Expression of the Emotions in Man and Animals*. Chicago: University of Chicago Press, 1965.

———. *The Origin of Species*. New York: Mentor, 1958.

Dawkins, Richard. *The Selfish Gene*. New York and Oxford: Oxford University Press, 1976.

———. "Selfish Genes in Race or Politics." *Nature* Vol. 289 (1981), p. 528.

Dewey, John. *The Influence of Darwinism on Philosophy*. Bloomington, IN: Indiana University Press, 1965.

Dobzhansky, Theodosius. *Genetics of the Evolutionary Process*. New York: Columbia University Press, 1970.

East, Edward M. *Heredity and Human Affairs*. New York: Scribner, 1927.

Eibl-Eibesfeldt, Irenäus. *The Biology of Peace and War*. New York: Viking, 1979.

———. *Ethology*, 2d ed. New York: Holt, Rinehart and Winston, 1975.

Eiseley, Loren. *Darwin's Century*. Garden City: Doubleday, Anchor, 1961.

———. *The Firmament of Time*. New York: Atheneum, 1966.

Evolution: A Scientific American Book. San Francisco: Freeman, 1978.

Fisher, Helen E. *The Sex Contract*. New York: Morrow, 1982.
Folsome, Clair Edwin. *The Origin of Life*. San Francisco: Freeman, 1979.
Fox, Robin. *Encounter with Anthropology*. New York: Dell, 1975.
Galton, Francis. *Hereditary Genius*. New York: Appleton, 1871.
———. *Inquiries into the Human Faculty and Its Development*. New York: AMS Press, 1973.
George, Henry. *Progress and Poverty*. New York: Modern Library, 1929.
Glaser, Hermann. *The Cultural Roots of National Socialism*. Austin, TX: University of Texas Press, 1978.
Goldman, Eric F. *Rendezvous with Destiny*. New York: Vintage, 1958.
Gould, James L. *Ethology*. New York: Norton, 1982.
Gould, Stephen Jay. *Ever Since Darwin*. New York: Norton, 1977.
———. *The Mismeasure of Man*. New York: Norton, 1981.
———. *Ontogeny and Phylogeny*. Cambridge, MA: Harvard University Press, 1977.
———. *The Panda's Thumb*. New York: Norton, 1980.
Gribbin, John. *Genesis*. New York: Delacorte, 1981.
Gruber, Howard E., and Barrett, Paul H. *Darwin on Man*. New York: Dutton, 1974.
Haller, Mark H. *Eugenics*. New Brunswick, NJ: Rutgers University Press, 1963.
Hampson, Norman. *The Enlightenment*. Harmondsworth, England: Penguin, 1968.
Harris, Marvin. *Cultural Materialism*. New York: Random House, Vintage, 1980.
Hatch, Elvin. *Theories of Man and Culture*. New York: Columbia University Press, 1973.
Hays, H.R. *From Ape to Angel*. New York: Knopf, 1965.
Himmelfarb, Gertrude. *Darwin and the Darwinian Revolution*. Garden City: Doubleday, Anchor, 1959.
Hinde, Robert A. "The Study of Temperament." *Nature* Vol. 293 (1981), pp. 607–8.
Hitler, Adolf. *Mein Kampf*. Boston: Houghton Mifflin, 1943.

Hofstadter, Richard. *Social Darwinism in American Thought*. New York: Braziller, 1959.

Holden, Constance. "The Politics of Paleoanthropology." *Science* Vol. 213 (1981), pp. 737–40.

Hrdy, Sarah Blaffer. *The Woman that Never Evolved*. Cambridge, MA: Harvard University Press, 1981.

Human Ancestors: Readings from Scientific American. San Francisco: Freeman, 1979.

Huxley, Julian. *Heredity East and West*. New York: Henry Schuman, 1949.

Huxley, Thomas H. *Man's Place in Nature*. Ann Arbor, MI: University of Michigan Press, 1959.

Irvine, William. *Apes, Angels and Victorians*. New York: McGraw-Hill, 1955.

Jeffreys, M.V.C. *John Locke*. London: Methuen, 1967.

Johanson, Donald, and Edey, Maitland. *Lucy*. New York: Simon and Schuster, 1981.

Jones, J.S. "How Different Are Human Races?" *Nature* Vol. 293 (1981), pp. 188–90.

———. "An Uncensored Page of Fossil History." *Nature* Vol. 293 (1981), pp. 427–28.

Joravsky, David. *The Lysenko Affair*. Cambridge, MA: Harvard University Press, 1970.

Kardiner, Abram, and Preble, Edward. *They Studied Man*. Cleveland: World Meridian, 1965.

Keith, Arthur. *The Antiquity of Man*. London: Williams and Norgate, 1925.

Kennedy, James G. *Herbert Spencer*. Boston: Twayne, 1978.

Kennedy, Kenneth A.R. *Neanderthal Man*. Minneapolis: Burgess, 1975.

Lamprecht, Sterling Power. *The Moral and Political Philosophy of John Locke*. New York: Russell & Russell, 1962.

Leach, Edmund. "Biology and Social Science: Wedding or Rape?" *Nature* Vol. 291 (1981), pp. 267–68.

Leakey, L. S. B. *By the Evidence*. New York: Harcourt Brace Jovanovich, 1974.

Leakey, L. S. B., and Goodall, Vanne Morris. *Unveiling Man's*

Origins. Cambridge, MA: Schenkman, 1969.

Leakey, Richard E., and Lewin, Roger. *Origins*. New York: Dutton, 1977.

———. *People of the Lake*. Garden City: Doubleday, 1978.

Lewin, Roger. "Do Jumping Genes Make Evolutionary Leaps?" *Science* Vol. 213 (1981), pp. 634–36.

———. "Evolutionary Theory Under Fire." *Science* Vol. 210 (1980), pp. 883–87.

Lewontin, Richard C. "Sleight of Hand." *The Sciences* Vol. 21, No. 6 (July/August, 1981), pp. 23–26.

Locke, John. *An Essay Concerning Human Understanding*. Oxford: Clarendon, 1975.

Lorenz, Konrad. *King Solomon's Ring*. New York: Signet, 1972.

———. *On Aggression*. New York: Harcourt Brace, 1966.

Lovejoy, C. Owen. "The Origin of Man." *Science* Vol. 211 (1981), pp. 341–50.

Lowie, Robert H. *The History of Ethnological Theory*. New York: Holt, Rinehart and Winston, 1937.

Lumsden, Charles J., and Wilson, Edward O. *Genes, Mind and Culture*. Cambridge, MA: Harvard University Press, 1981.

———. "Genes, Mind and Ideology." *The Sciences* Vol. 21, No. 9 (November 1981), pp. 6–8.

Lysenko, Trofim Denisovich. *Agrobiology*. Moscow: Foreign Languages Publishing House, 1954.

Martin, Kingsley. *The Rise of French Liberal Thought*. New York: New York University Press, 1954.

Mayr, Ernst. *Populations, Species and Evolution*. Cambridge, MA: Harvard University Press, Belknap, 1970.

Mead, Margaret. *Sex and Temperament in Three Primitive Societies*. New York: Morrow, 1963.

Medvedev, Zhores A. *The Rise and Fall of T. D. Lysenko*. New York: Columbia University Press, 1969.

Mellen, Sydney L.W. *The Evolution of Love*. San Francisco: Freeman, 1981.

Millar, Ronald. *The Piltdown Men*. New York: St. Martin's, 1972.

Miller, William. *A New History of the United States*. New York:

Dell, 1968.
Mind and Behavior: Readings from Scientific American. San Francisco: Freeman, 1980.
Montagu, Ashley. *The Nature of Human Aggression*. Oxford: Oxford University Press, 1976.
———, ed. *Sociobiology Examined*. Oxford: Oxford University Press, 1980.
Morris, Desmond. *The Human Zoo*. New York: Dell, 1971.
———. *The Naked Ape*. New York: Dell, 1969.
Mosse, George L. *The Crisis of German Ideology*. New York: Schocken, 1981.
———. *Toward the Final Solution*. New York: Harper & Row, Colophon, 1980.
"Myths, Models and Human Evolution." *Nature* Vol. 287 (1980), p. 385.
Pastore, Nicholas. *The Nature-Nurture Controversy*. New York: King's Crown, 1949.
Peel, J.D.Y. *Herbert Spencer*. New York: Basic Books, 1971.
Plomin, Robert; DeFried, J.C.; and McClearn, G.E. *Behavioral Genetics*. San Francisco: Freeman, 1980.
Poliakov, Leon. *The Aryan Myth*. New York: Basic Books, 1974.
Reader, John. *Missing Links*. Boston: Little, Brown, 1981.
Rensberger, Boyce. "Facing the Past." *Science 81* Vol. 2, No. 8 (October 1980), pp. 40–51.
Rose, Steven. "Genes and Race." *Nature* Vol. 289 (1981), p. 335.
Rosenberg, Alfred. *Race and Race History*. New York: Harper & Row, 1971.
Ruse, Michael. *The Darwinian Revolution*. Chicago: University of Chicago Press, 1979.
Russell, Bertrand. *A History of Western Philosophy*. New York: Simon and Schuster, 1945.
Seger, Jon. "Social Scientists and Sociobiologists Get Their Lines Crossed." *Nature* Vol. 291 (1981), p. 690.
Shattuck, Roger. *The Forbidden Experiment*. New York: Farrar, Straus and Giroux, 1980.
Siemens, Hermann W. *Race Hygiene and Heredity*. New York:

Appleton, 1924.

Singer, Peter. *The Expanding Circle*. New York: Farrar, Straus and Giroux, 1981.

Skinner, B.F. *Beyond Freedom and Dignity*. New York: Bantam/Vintage, 1972.

———. *Reflections on Behaviorism and Society*. Englewood Cliffs, NJ: Prentice-Hall, 1978.

———. *Science and Human Behavior*. New York: Free Press, 1965.

Spencer, Herbert. *Essays*. New York: Appleton, 1914.

———. *First Principles*. London: Watts, 1937.

———. *Social Statics*. New York: Robert Schalkenbach Foundation, 1954.

———. *The Study of Sociology*. New York: Appleton, 1908.

Stanley, Steven M. *The New Evolutionary Timetable*. New York: Basic Books, 1981.

Stebbins, G. Ledyard. *Processes of Organic Evolution*, 3d ed. Englewood Cliffs, NJ: Prentice-Hall, 1977.

Tenenbaum, Joseph. *Race and Reich*. New York: Twayne, 1956.

Thomson, Robert. *The Pelican History of Psychology*. Harmondsworth, England: Penguin, 1968.

Tiger, Lionel. *Men in Groups*. New York: Random House, Vintage, 1970.

———. *Optimism*. New York: Simon and Schuster, 1980.

Tiger, Lionel, and Fox, Robin. *The Imperial Animal*. New York: Dell, 1974.

Trevelyan, G.M. *History of England*, Vol. 2. Garden City: Doubleday, Anchor, 1953.

Veblen, Thorstein. *The Theory of the Leisure Class*. Harmondsworth, England: Penguin, 1979.

Viereck, Peter. *Metapolitics from the Romantics to Hitler*. New York: Knopf, 1941.

Wade, Nicholas. "Voice from the Dead Names New Suspect for Piltdown Hoax." *Science* Vol. 202 (1978), p. 1062.

Wall, Joseph Frazier. *Andrew Carnegie*. New York: Oxford University Press, 1970.

Washburn, S. L. "The Piltdown Hoax: Piltdown 2." *Science* Vol.

203 (1979), pp. 956–98.

Watson, John B. *Behaviorism*. New York: Norton, 1970.

Weiss, Mark L., and Mann, Alan E. *Human Biology and Behavior*, 2d ed. Boston: Little Brown, 1978.

Williams, B.J. *Evolution and Human Origins*, 2d ed. New York: Harper & Row, 1979.

Williamson, Peter G. "Morphological Stasis and Developmental Constraint: Real Problems for Neo-Darwinism." *Nature* Vol. 294 (1981), pp. 214–15.

———. "Paleontological Documentation of Speciation in Cenozoic Molluscs from Turkana Basin." *Nature* Vol. 293 (1981), pp. 437–43.

Wilson, Edward O. "Genes and Racism." *Nature* Vol. 289 (1981), p. 627.

———. *On Human Nature*. Cambridge, MA: Harvard University Press, 1978.

———. *Sociobiology: The New Synthesis*. Cambridge, MA: Harvard University Press, Belknap, 1975.

Yolton, John W. *Locke and the Compass of Human Understanding*. Cambridge: Cambridge University Press, 1970.

Young, J.Z. *Programs of the Brain*. Oxford: Oxford University Press, 1978.

Zihlman, Adrienne L. "The Sexes Look at Sex." *Nature* Vol. 294 (1981), pp. 42–43.

Zinn, Howard. *A People's History of the United States*. New York: Harper & Row, 1980.

Index